AFRICAN NIGHTS

ALSO BY KUKI GALLMANN

I Dreamed of Africa

AFRICAN NIGHTS

KUKI GALLMANN

TRUE STORIES FROM THE AUTHOR OF
I Dreamed of Africa

Perennial
An Imprint of HarperCollinsPublishers

Designed by Stanley S. Drate/Folio Graphics Co. Inc.

Library of Congress Cataloging-in-Publication Data
Gallmann, Kuki.
   [Night of the lions]
   African nights : true stories from the author of I
dreamed of Africa / Kuki Gallmann.
      p. cm.
   Previously issued as two works: Night of the
lions. London, England : Viking ; New York : Penguin
Putnam, 1999; and African nights. London : Viking,
1994.
   ISBN 0-06-095483-3
   1. Kenya — Social life and customs. 2. Laikipia
District (Kenya) — Social life and customs.
3. Gallmann, Kuki. I. Title
DT433.54.G35   2000
967.62—dc21   99-44291

08 09 10 ❖/RRD 20 19 18 17 16 15 14 13 12

For Sveva, a child of Africa,
with my love forever.

Listen attentively, and above all remember
that true tales are meant to be transmitted; to
keep them to oneself is to betray them.

BAAL-SCHEM TOV (ISRAEL BE ELIEZER)

Both loved stories,
And when I tell mine, I hear their voices.
Whispering from beyond the silenced storm
They are what links the survivor to their memory.

ELIE WIESEL, Souls on Fire

Contents

Part III

CREATURES FOUND AND LOST

Part IV

A WORLD BEYOND

Part V
A TASTE FOR ADVENTURE

Photographs follow page 108.

Introduction

Dreams can change your life, and eventually the world.
FATIMA MERNISSI, *The Harem Within*

When I first came to Africa, the people addressed me as Memsaab. In time, they baptized me Nyawera, The One Who Works Hard. Now they call me Mama, because I have chosen to stay, and because I belong.

I was born in Italy, and from earliest childhood I dreamed of Africa. After recovering from a tragic accident which left me crippled, I went to live in Kenya with my second husband, Paolo, his two daughters and the son from my first marriage, Emanuele, then six years old. We bought a house in Nairobi, in an area called Gigiri, and spent much time exploring the country, particularly the coast, where we used to spend most of our holidays. After many adventures, we acquired Ol Ari Nyiro, a vast ranch on the Laikipia plateau overlooking the Great Rift Valley, which became our home.

The ranch was magnificent and abounded with wildlife. We managed it to keep a balance between agriculture, livestock and the conservation of the environment, in those happy, unforgettable days. We learned to know Africa, to love its people and to protect its wild animals and plants.

In 1980, a few months before the birth of our child, Paolo was killed in a car accident while bringing me her crib, a wooden boat made out of a tree trunk by fishermen at the coast. Our daughter Sveva was born beautiful and the image of her father. I chose to remain in Laikipia. Here Emanuele, who since early childhood had shown an unusual intelligence and maturity beyond his years, developed a deep passion for snakes. At seventeen he was killed by one of his vipers while extracting venom for the manufacture of an antivenin. He died in my arms in a few minutes. I buried him next to Paolo, at the foot of my garden, and planted on each grave a yellow fever tree in a sign of survival and hope. In my despair I looked around me, and I saw Africa.

Africa is a continent of extremes.

There are droughts and there are floods. There is an Africa of tragedy and famine, of corruption and war, of blood and hunger and tears, of incurable disease and tribal clashes and misery and violence and political unrest. It is the Africa we read about today in every paper, the one we see daily in biased cable television reports. It is an Africa captive to and dependent on the blackmail of foreign aid, constantly judged, constantly criticized and never understood. Here the rich West has imprinted its competitive, frantic image, created alien needs, imposed alien philosophies and financed impossible schemes, unsuited to the potential and true spirit of this troubled and fantastic continent, all too ready to take back that help and sit in judgment of yet another failure.

I do not sing that Africa. There is no need for another negative reportage, which will leave a bitter taste and serve no purpose.

There is a different side to this ancient land. It is the Africa that, since the beginning of time, has evoked in travelers a deep recognition, an inexplicable yearning to return. The place that still has what most of the world has lost. Space. Roots. Traditions. Stunning beauty. True wilderness. Rare animals. Extraordinary people. The land that will always attract those who can still dream.

This Africa, into whose embrace went Paolo and Emanuele,

called early to the land of beyond, in turn healed my sorrow and became my life purpose. This maternal, primordial Africa taught me acceptance, endurance and survival. I recognized it as a place to find wisdom. A place to end this journey and begin a new one. A place of renewal and rebirth. A good place to die.

In the very heart of Kenya where I still live—Ol Ari Nyiro, The Place of Dark Waters—I walked the hills and the valleys in those dark days of solitude, tiring my body, asking the wind the questions whose answers are the reasons for life. I missed Paolo, a man for all seasons and places, handsome, gifted with irresistible charm, eloquent and brave, happy in the bush and at ease in a palace; I missed his poetry and flair. I missed Emanuele, who taught me first what it means to be a mother, my sensitive, intelligent boy, whose eyes had the sadness and wisdom of ages, who was taken into that good night by one of the snakes he loved.

Walking alone through the pervading magic of untouched African landscapes, open to growth as one is when at the bottom of pain, I felt quintessentially part of the whole. One evening, looking down at the breathtaking depths of the Mukutan Gorge in the Great Rift Valley, in this living cathedral of the spirit I discovered my crusade and found peace. There was Sveva, the daughter Paolo would never know, the radiant girl who was my hope and my future. I was in Africa, and this was my cure. In memory of Paolo and Emanuele I started the Gallmann Memorial Foundation and devoted the rest my life to making a positive difference for the African environment.

When I sit alone at night at my desk that looks down over the Savannah or with my daughter in front of our fireplace at Kuti, and the peculiar silence of the African night is full of the voices of a million crickets and interrupted by a faraway hyena, by a lion's roar or by our dogs barking suddenly at the shadows of elephants, I remember the adventures of the past, the people I have known, the men we have lost and I like to tell their stories.

Mine are love stories about Kenya, my Africa. It is the Africa of sunshine and endless vistas, of roaming herds on the plains, of red dust and galloping giraffe, of forests and snow and prehistoric lakes, of gentle, handsome, intelligent people whose poverty is not of the spirit. People for whom tradition is important and to whom family values still matter; people who protect the young and respect the old, care for the sick and feed the hungry, even if it means sharing the little that they have; generous people, ready to smile and to forgive; people with a song in their heart and a dance in their step; enduring, compassionate and infinitely patient. The people of Kenya, whose ancient, proven wisdom I respect. I salute them and I thank them for having allowed me to live amongst them, to bury my men in the soil of my garden, as Africans do, and for allowing me the honor and the choice of becoming a Kenyan like them.

I have new stories to tell. I have old stories that I have not yet told.

The elephant raid my garden at Kuti after dark as they did a quarter of a century ago, I encounter—now more than ever—buffalo in my car lights, there are still rhino coming to the salt lick below the Nest and in the moonless nights the lion prey on the cattle around the boma as they have since the beginning of lion and cattle in Africa. Butterflies and rare birds bejewel the forests and on the silent lakes millions of pink flamingo forage. The Pokot herders stand on one leg, a black blanket thrown on a shoulder with supreme elegance, necks bright with beads and an ostrich feather beckoning on their skullcap. The waves of another tide reach silently along the mangroves, while the fish eagle surveys her hunting grounds somewhere on the north coast, off the beaten track. There are still remote places untouched by human progress, where harmony prevails.

Magic, and the inexplicable, still touch our lives. This is the lyrical, therapeutic Africa that I describe, the one I live every day of my life, the one that it is my life aim to preserve.

Scores of bright young Kenyans visit the Wilderness Education

Center that I built to honor the memory of my son, and discover in Ol Ari Nyiro their Africa. Each of these children is a messenger, who will carry this crusade a step further. Through education and experience the generation of today will realize the importance of their unique environment. For the world needs beauty and youth needs hope.

It is to these young people and to my daughter Sveva, whom they call Makena, that I dedicate this collection of true stories.

KUKI GALLMANN
Laikipia, 1999

The Gallmann Memorial Foundation
P.O. Box 45593
Gigin Road
Nairobi
Kenya
http://www.gallmannkenya.org

Acknowledgments

I would like to thank my editor, Larry Ashmead at HarperCollins, for his competence and for cleverly rearranging my stories in this new order; Allison McCabe at HarperCollins for her enthusiastic support; Toby Eady, for his professional and brotherly help; Gilfrid Powys for his total confidence in me; all my African friends at Ol Ari Nyiro, particularly Patrick Ali, Simon Itot, Silas Kaboto and Issak Golicha, for holding down the fort when I was deep in my writing; Benjamin Woodley, Murray Grant and Shahar Gelbart, the closest amongst my "adopted sons," for their love and loyalty and the fun we have had; my readers, who have asked me to go on writing my stories, who have come to Africa to discover their truth and who, having found it, have become the goodwill ambassadors that this continent needs. And my daughter Sveva, for her love, her spirit and her smile.

Part I

OFFERINGS

A Maasai Woman

In the face of some Masai matriarchs could be read the tale
of a people whose iron code of tradition makes them
unique among the earth's beings.

ROBERT VAVRA, *A Tent with a View*

The woman who came through the camp was lean and tall.
She could have been of any age between eighteen and thirty.
She marched straight towards me in the yellow August dawn,
while I stretched to chase away the shadows of sleep, shivering in
the early-morning air of the coldest month of the year on the
Kenya Highlands. It was 1973, when hunting was still allowed in
Kenya.

She greeted me in Swahili and in a high clear voice, without
any shyness, she asked me immediately for salt.

'*Chumvi. Mimi nataka chumvi.*' She smiled with even, well-
spaced teeth.

All creatures in the Highlands need salt to supplement their
diet. Rock salt mixed with the soil creates a salt lick irresistible to
elephant, rhino, antelope and buffalo. They walk long miles at
dusk, drawn by its subtle scent, imperceptible to human nostril.
But before leaving the shelter of the shrubs around the area of
the salt-lick, which generations of converging animals' hooves
have made barren of vegetation, they pause and sniff the air with
quivering muzzles, with tentative trunks, to detect any smell of
danger in the wind. Reassured, they move on, head down, eager
to lick the salt trapped in the earth.

Chumvi. A handful of the precious salt is a treat that few humans, even, can resist in wild Africa.

I smiled up at her, and nodded. She came close on elegant legs, and sat in the dust next to me.

We had camped in the late afternoon, not far from a *manyatta* in the area of Narok, one of the main centres of the proud Maasai tribe. It then consisted of a couple of petrol stations, a general store kept, like most other stores, by Indian merchants, and a few primitive *dukas,* shops where one can find a bit of everything, from tea to blankets, from dark sugar-cane to snuff, from tinned beans to tablets to fight – often in vain – the endemic malaria.

We had chosen a spot in the shade of some yellow fever trees, and at nightfall had lit a fire of sticks and dry branches. There we had barbecued, on some rudimentary iron wire, the tender fillet of a Thomson's gazelle that had not been fast enough.

The *manyatta* was a large one, composed, like all others, of longish low huts rounded like loaves. Made from a mixture of mud and dung plastered on a frame of curved sticks, they reminded me of dried-out chrysalises. The huts were surrounded by a thick barricade of acacia and 'wait-a-bit' thorn branches, arranged so that the spikes were impenetrable by animal or man.

Cattle are the wealth that the god Ngai – the sky in Maasai – had bestowed forever on the Maasai race, and it was within this enclosure that their livestock spent the night, each animal packed close to the other, sheltered from predators and cattle rustlers.

The woman was dressed in goatskins, reddened with fat and ochre. From her right ear, stretched down to her shoulder, hung a tin ornament polished like silver and shaped like an arrow. Her left ear was studded with an old beer bottle top, shiny as a new coin.

Her right leg, from ankle to knee, was encased in a spiralling brass bracelet, so tight that it scarcely allowed any space for her slim bird-like leg. This was so skeletal that it reminded me of the ancient Roman shinbones I had once found as a child in a newly

ploughed field I was exploring with my father in the countryside around Quarto d'Altino. How far from Africa my childhood seemed, yet how close.

Her ornaments showed that this woman was married. Innumerable rows of small coloured beads swayed gently from her neck, on her forehead, and around her stunning black eyes, like a dancing mask. They were threaded with infinite patience, extravagant skill and a symmetrical elegance that no mirror had suggested. They framed her tilted eyes, around which countless flies, motionless and undisturbed, formed dainty patterns like *points d'esprit* on the lacy veil of an Edwardian bonnet.

I poured the salt straight from the plastic bag onto her pink palm, proffered as a cup held between her slender black fingers. She licked it laughing, greedily, like the most sought after delicacy. Only when she had finished, and I had put the remainder of the packet in her hand, did she look me straight in the face and begin to ask questions. Because she spoke Swahili, which was unusual in those days for a Highland Maasai, I could understand her.

Our talk was oddly feminine and for a time we became as close as one can only be over a handful of salt in the solitude of a newborn day, when the Maasai men, carrying their spears against lions and thieves, had gone out grazing their herds among far-off lowings, and the European men, in a hurry to follow fresh buffalo tracks, had forgotten to drink all their tea.

The only sign left of Paolo that morning was a still steaming cup, and the squashed stub of his first cigarette. Quick to notice these traces of a male presence, '*Wapi bwana yako*?' she asked.

'Where is your husband?' But before I could answer about mine, she told me of hers. Her story was typical of all the young beauties of her tribe and her age-group, just after circumcision.

A handsome *moran*, or young warrior, who had already won, by raiding, enough cattle to afford a wife, had shown his interest by offering her a necklace, which she had accepted. Her parents then waited in anticipation for her betrothed to come, according to custom, with his first gift of honey. This, she and the girls circumcised with her – and because of this forever her sisters – mixed with milk and then drank together.

Soon after, vast quantities of honey had been offered by the future groom, and fermented and distilled into a heavy liquor that the elders drank amidst celebration. At this stage the young man had been summoned, and was told – may God listen – that his ritual gifts had been accepted and his request granted. No one else might now come to claim his bride.

Her old mother had received a lamb, and her father a calf and cured sheepskins with which to prepare the wedding garment. Two calves and a bull, all white like the cool new moon, had been brought by the groom on the wedding day. As prescribed by custom, they had to be healthy and strong, without marks to scar the soft sun-warmed hides. The unfortunate sacrificial victims of African celebrations – a ram and two hoggets – had been slaughtered, and the wretched ram's fat used as a ceremonial ointment.

On the given day, after the sun had risen like an incandescent *calabash* on the indigo horizon, her friends sang a high-pitched song of joy, her heart leapt like an impala in the happiness of her special ceremony, and the old women had come. They had doused her with that brew of honey and, helped by her proud mother, they had fixed the ornaments to her leg and ears with elaborate ritual.

I listened, fascinated. The story came in bouts, prompted by questions, and facilitated by half a pound of sugar-cane that she licked casually and with glee, straight from the bag. The sun rose high, and the cicadas' song grew deafening, dry like the sound of twigs beaten together by a thousand hidden hands.

The tall slim woman yawned and stretched. I understood that for today her story was finished. She looked around and her eyes focused on the green cake of soap, drying on a stone, with which

Paolo had washed his hands. She pointed at it with a sudden jerk of her chin, emphasizing her desire by pretending to spread it onto her arm and smelling her skin with a beatific smile. I passed it to her and she began to smooth it all over her dry skin like a cream, moaning with pleasure. The flies, uncaring, flew dozily from the restless eyelids, and settled back there immediately.

I asked how many children she had borne. She thought a bit about it, and finally opened three fingers, but, as an afterthought, she slapped her stomach energetically:

'*Mimi ni mimba tena,*' she announced proudly, with a toss of her head. 'I am pregnant again.'

In her voice was a ring, like a fresh bell at dawn.

I was surprised by her thinness and I told her so. She explained that the Maasai were careful not to let pregnant mothers get too fat, as this was regarded as dangerous for the baby. How modern. But pregnant women had the unique privilege of eating meat: an astonishing luxury for this tribe which feed purely on the blood, urine and curdled milk of their cattle. Only stolen meat is normally eaten, never that of their own livestock.

She rose with one fluid movement, and measured me with a cocky, challenging look in her eyes. Suddenly curious she asked:

'*Bwana yako ulilipa ngombe na njan ngapi kwa baba yako kuolewa wewe?*' 'How many cows and calves did your bwana pay your father for you?'

Slightly humiliated by the inadequacy of our European traditions, I tried to explain that in the land called Ulaia, where I came from, we followed other *desturi. Desturi* means custom in Swahili, and is a magic word to unfold the inexplicable. Customs are sacred, unquestionable, instantly accepted without reserve. Often, hiding behind the excuse of my *desturi* had saved me from potentially embarrassing explanations.

It was clear that the woman considered this particular *desturi* undignified, and for a moment it looked so to me also. She made no comment, just lifted her shoulder, almost inperceptibly, to dismiss that incomprehensible European meanness.

With a tinkling of anklets she stood tall, light in her thinness, without leaving any mark on the dust where we had sat together. Then, with a natural grace, she parted the skin garment which, like a peplum, covered her chest, and exposed two pendulous breasts, swollen like ripe oblong gourds. With thumb and index she lifted one, and squeezed it with the expert gesture of the milker. A long opalescent spray spurted out, darting inches from my face, hitting the bushes with a sharp noise.

With a proud jerk of her head she invited me to do the same. But before I could admit my defeat, I could detect in her frank, laughing stare a teasing note. A sudden breeze ran unexpected through the treetops, touched our faces and was soon forgotten.

She went away amongst the low sage bushes, without a word, her head held high, as when she had first appeared.

Mwtua

Dis aliter visum.*
Virgil, *Aeneid*, II, 428

He was a little man with a perennial grin. His short greying hair, small eyes brightened by a continuous smile, his readiness to obey or volunteer for any work and his intrinsic innocence, were like a peasant-saint's in biblical tales. He had been with me for many years, looking after my house in Nairobi, a reliable fellow respected by all, kind to children and adored by dogs. He was not intelligent, on the contrary perhaps a bit simple, and his sentences often became tangled in a painful stutter, but his good nature and willingness amply made up for his lack of initiative.

I had noticed that recently he had been looking old and become rather forgetful. His stammer had increased, making it more painful for him to answer quickly, and harder for me to understand what he wanted to say. He ironed perfectly but his eyesight seemed to be failing him, and often I found strange garments mixed up with mine and unknown pullovers amongst my shirts.

He looked tired and he shuffled. I began to wonder if he should not retire, and go back to Kitui where he had come from,

---

*'The gods thought otherwise.'

to look after his grandchildren, and the small *shamba* he had acquired over the years.

He did not want to go; and as if sensing that I was about to call him to explain that the time had arrived for him to return home and retire in peace, he seemed to double his efforts, to work more and longer hours, as if he wanted to prove to me that, in fact, he still had many years left of active work.

One night, coming back late from a dinner party to my house in Gigiri, instead of the usual guard, a strange little man, trussed up in a too-large nightwatchman's greatcoat, trotted up to the gate and fidgeted endlessly with keys before he finally managed to open it. The oversized helmet on the small grey head had slipped down almost to cover his eyes, but revealed a happy, slightly fanatic grin: it was Mwtua. The *askari* had been taken ill by a sudden attack of malaria, so he had volunteered to help and was spending all night up in the chill, faithfully guarding my house.

Despite all this, I realized that Mwtua had to go, but I wanted to find the right opportunity to tell him.

It was a foggy morning in Nairobi, and when I tried to talk to my ranch on the Laikipia Security network, my radio did not seem to be receiving properly. It had been raining heavily the night before and I wondered if the unusual amount of static meant that a branch had fallen on the aerial. I called Wangari, the maid who was Mwtua's niece, and asked her to go and find a gardener to discover if the aerial was still in the upright position. She was away a few minutes.

'*Ndio,*' she explained. '*Aerial naaunguka, lakini Mwtua nasema yeye nawesa kutanganesa.*' 'Yes. The aerial has fallen, but Mwtua said he can fix it.'

I smiled. It was so typical of him. It was naturally out of the question. The aerial was very high, tied to the top of the tallest tree. The radio people from Wilken had come with a special ladder to do it. It was impossible to reach it otherwise.

'I shall call the maintenance people,' I said to Wangari. 'Please tell Mwtua we will take care of it.'

The telephone was ringing, so I went to answer it. Then I tried to call the radio workshop, but the line was engaged.

I noticed that in the meantime it had started drizzling again. I went to look out of my window. A short wooden ladder was leaning against the Cape chestnut in the middle of the lawn, to which the aerial had been fixed.

A ladder? Why? With a sudden premonition, I reached for my glasses. I looked. Sure enough, amongst the leafy branches, in his green uniform and practically camouflaged, Mwtua was climbing, agile and fast, towards the aerial. I caught my breath: this was impossible. The branches were so thin towards the top, surely they would not support a human body.

The tribesman of Kisii and Mkamba are forest people; they love working with wood and they know trees. As children they learn to climb for fruit and honey or to steal birds' eggs, while they tend the goats and cattle in the forests. Still, that tree was too high and wet with rain, its branches swaying, unsafe. Mwtua was too old to go climbing. I was about to open the window to call him down, when something about him made me stop.

It was as if a change had come upon him on that tree. His old man's movements had been shed like old skin. A young Mwtua was climbing, alert and nimble, with soft fluid gestures. His thin legs and arms seemed to wrap themselves easily around the branches with a prehensile skill, inherited from bygone arboreal generations. But what was most extraordinary was the transformation of his eyes. They were open and enlarged so that the white part seemed huge, almost phosphorescent, and their still gaze was fixed like an animal's. He reminded me uncannily of a bushbaby I had once kept.

I stopped breathless, looking at him mesmerized.

I did not know that downstairs our *ayah*, Wanjiru, was looking up to Mwtua from the kitchen window with the same apprehension. She told me later that her impressions of Mwtua were identical to mine.

Something had definitely happened to him up in that tree.

So deep seemed his transformation, and so remote was he, absorbed in his world of leaves and air, that I was afraid to startle him by opening the window. I decided to attract his attention by rapping on the glass instead, but he did not seem to hear. Then he looked up, like a bird surprised by a strange sound. At that moment the telephone rang in my room again and I went to answer. I was replacing the receiver when I next looked out.

It was raining heavily now, and through the rain I saw the treetop oscillating. Then, in front of my eyes, to my eternal horror, Mwtua's lithe body precipitated, with the slow motion of a nightmare, head first in a shower of leaves, no different from a leaf himself, on to the lawn, to lie there, motionless. A small broken branch fell with him and the radio cable, like a useless liana, swung forlorn in the air. I opened the window to see better.

He lay in a crumpled heap, pathetically small in his green clothes and, with a lump in my throat, I was sure he was dead. No one could have fallen on his head from that height without breaking his neck.

After what seemed a very long stillness, there was a sudden flutter of activity. Like spectators invading the stage after the show has ended, the gardener, the *shamba* woman and Wanjiru ran towards Mwtua as if they had all been waiting to spring into action. At the same time Wangari put her apron over her head, threw up her hands to the sky and, in a new wild voice which sent a chill down my spine, started wailing in an unknown language an eerie and ancient song of mourning.

Wanjiru was on him. I noticed she had kicked her shoes off to run faster.

'Don't touch him!' I screamed from the window, afraid that unskilled handling would make him worse if, by some quirk of fate, he was still alive.

'Kwisha kufa?' 'Is he dead?' I called out, praying that this night-mare would finish, hoping to turn back the clock.

'Badu!' 'Not yet!' Wanjiru screamed back.

The doctor, then.

I realized that every second could be crucial, and that the right action could make all the difference. With flying fingers, I dialled the home number of the Italian brain-surgeon, a great friend, whom I contact in any emergency. It was a short cut to avoid lengthy explanations to a telephone operator.

'Marieke,' I begged his wife, 'Mwtua has fallen from a tree and I think he is dying. Please tell Renato I am bringing him to Nai-robi Hospital now.'

A former nurse herself, she did not waste time with idle ques-tions. I slammed the phone down and ran downstairs.

Surrounded by moaning people, Mwtua was curled up in the foetal position, with his eyes shut. Some mown grass was stuck to his cheek. He looked quite dead.

I felt the great inner silence which anticipates irreversible doom, and in this soundless world I knelt at his side. I forced myself to open one of his eyelids – the skin was cold and clammy – and touched the pupil lightly with a leaf. To my overwhelming relief, the eye contracted, flickered. A shiver ran through his body: he was alive!

Sound crept back in my consciousness, and I became aware of a rasping breathing that came from his chest in bursts. I put my hand on his back and massaged him, calling him softly. Pearly saliva, mixed with grass fragments, frothed from his lips.

We drove him to hospital in the back of my car, wrapped in a blanket, writhing but unconscious, and jerking in his sleep as if he dreamt he was still climbing the tree.

He was put straight into the Intensive Care unit, and the hos-pital machinery organized by Renato Ruberti began to hum effi-ciently around him. Blood pressure, temperature, X-rays, physical tests, scans and all kinds of examinations were performed quickly and smoothly. Then Renato looked up at me, while still holding Mwtua's wrist. For a long moment his intelligent eyes behind his spectacles stared into mine before he spoke.

I swallowed.

A broken neck? A fractured skull? A smashed thorax? An irreversible coma? Perennial brain damage? Death in a few minutes?

A sudden grin split his face in two.

'You will not believe it,' he drawled in Italian. 'He has nothing wrong with him at all. Only a slight concussion. The tree was thirty metres tall you said?' He shook his head. 'Not a cracked rib, and not a scratch. He does not even need a plaster.'

The staff howled, accepted his revival as an act of magic and praised God – who decides who should live and who should die – because in his wisdom he had spared Mwtua, who knew no evil. Masses were celebrated at the mission church back in his village, and special tribal ceremonies of thanksgiving, so that the new and old gods might be appeased.

As a precaution I kept him for a week – during which he mostly slept and was woken regularly only to be hand-fed – under observation in the Intensive Care unit and Special Attention ward at the hospital.

After that a stream of people came to see him. The visitors looked at him in silence and awe, a respect and consideration reserved only for a shaman or *muganga:* they declared there had been a miracle.

In their stories the tree became taller and taller, a holy force had lifted Mwtua up, a bird had delayed his fall. His adventure gained colours and new details every time it was repeated, as it is normal with legends.

Wanjiru declared that Mwtua had survived because God loved me, and would not allow a tragedy like this to stain my *boma.* Amongst most Kenyan tribes, the people about to pass away were traditionally taken outside their compound, as a house where someone had died was considered impure and should normally be burnt.

Although he had remained inexplicably unhurt, something had happened to Mwtua's head. He wandered around absently, with a beatific expression painted on his face, and nothing seemed to move him. He smiled more than ever, muttered to himself, played with the new puppies a great deal and sat outside his house in the staff quarters, in a sort of inertia. Tended by his wife, who had come to look after him from the village, he just stared ahead as if contemplating mysterious images within himself.

The doctors suggested that he should do some easy work, and Wanjiru allotted him simple repetitive tasks, like cleaning silver or shoes. He agreed with great enthusiasm, but he held the shoes upside down, polishing the soles, and we soon gave up.

Mwtua went back to his village in the end, where he looked forward to playing with his grandchildren and resting in his *shamba* as becomes an elder.

I was sorry to part with him. But when he came, dressed in a coat which had been Paolo's, to say his long, laborious farewells, shaking my hands and Sveva's again and again as if he did not want to let go, and promising he would come back as soon as possible, we noticed, with amazement, that he had completely lost his stammer.

Langat

When we build, let us think that we build forever.
JOHN RUSKIN, *The Seven Lamps of Architecture*

A woman approached on the road at noon, old, bent, dressed in rags. She walked up to me with slow shuffling steps, bare knotted feet covered in dust. It was dry and hot at Kuti. Again we had missed the rains.

I looked at her: haggard face, colourless handkerchief tied around her grey head. Long earlobes, all her teeth virtually missing. Yet perhaps she was not that old. She greeted me. I answered and waited.

'I am hungry,' she said in halting Swahili. 'I come from far.'

'Who are you?' I reached into my pocket.

A vague reminiscence stirred, of eyes which were once brighter, a lost twinkle, a pride, the dignity of being married to a good man.

'I was Langat's wife.'

The memories came back, like running children.

When we decided to build the house, Paolo called up the *fundis*, and I went to meet them.

'Here is Arap Langat,' he said; and I found myself looking into clear, grey and unblinking wise eyes.

Langat was short and almost plump. His cropped hair was very

16

white, his dark face round, with a small nose, wide cheekbones and even white teeth untarnished by age and chewing tobacco. His earlobes had been stretched – in Nandi custom – and hung down to his shoulders.

What struck me most, apart from his unwavering eyes, which looked straight, piercingly into mine, as if to measure me up, was his air of self-assured dignity, and the composure which emanated from him. I learned later that this came from the knowledge of his competence and the versatility in his job, which he liked and in which he took great pride.

Nguare and Lwokwolognei were his assistants. Nguare was a middle-aged Kikuyu, whose specialty was woodwork. His name – inappropriately, as he was a quiet, serious man with slow feet, usually awkwardly wrapped in ill-fitting clothes – was the Swahili word for francolin, an alert, scuttling little creature, always dashing into the shrubs at the side of the road. Extremely slow and precise in his work, Nguare had the peculiar habit of repeating always the last word of a phrase. He had a pale brown-yellow face in great contrast with the deep ink-black of Arap Langat and of Lwokwolognei.

Often one finds in an African features that are startlingly similar – but for the shade of his skin – to a European equivalent. Nguare was the practically identical brown version of a long-lost friend in Veneto, Alvise: and the African's smiling face never failed to remind me of that other, surrounded in the dreaminess of my memory by the drifting fogs of the Laguna.

The trio was completed by Lwokwolognei, junior, and still an apprentice at that time. With the vertiginous thinness found only in Turkana, he had a lean lustrous face, in which the only eye shone with a doubly vivid light, as if to compensate for the other, lost we never knew how. The eyelid perennially closed over the empty socket, gave him, when seen in profile, the melancholic look of a secretary bird. Lwokwolognei did a bit of everything, from masonry to carpentry, but his real passion was wood carving, at which he excelled, displaying in it a rare imagination and a naive artistry.

He had a bright young wife called Mary, industrious in embroidering goatskins with intricate patterns of beads and shells. But one day she died in childbirth and Lwokwolognei was seldom seen to smile again.

We had moved from Italy to Kenya a few years before and only recently acquired this vast piece of land in the Highlands, on the edge of the Great Rift Valley. It was still early days here in Laikipia, when everything had yet to be learnt, yet to be built, and we were trying to give physical substance to the shape of our dreams.

Constructing our house was one of the first steps, and as Paolo, in this new venture, had so much else to attend to, the task of deciding what we needed and of supervising its construction was left largely to me.

I looked around and found the objects that nature had created and left, like a generous, careless artist, exposed for us to discover and to enjoy. I loved to use these local materials, rocks from the river and boulders from the hills, logs cut from the forest's red cedar, or old twisted olive trunks, bleached by generations of suns and sculpted into fantastic forms by the insuperable art of strong winds.

Langat knew what he could achieve and was able to put into practice my architectural whims. He let his dormant, instinctive, tribal fantasy prevail, yet blend with the stereotyped European building notions that had been hammered into him in the days of his apprenticeship. In this combination of inspired intuition and acquired skill lay the roots of his excellence.

He was infallible in finding exactly what I wanted. After the first few times, when we had driven out together, looking for a special shape in the stones or grain in the trees, he was quick to pick up and interpret this new European extravagance: the preference for rough rocks to neat smooth cement; for old twisted *mutamayo*, forgotten by termites or too hard for them to devour, to planks of sawn factory timber; for grass thatch and palm roofing, reminders of breezes and savannah, instead of the shining corrugated iron sheets which today scar the African plains.

We began to build the main house with its sitting room and verandah. When some difficult spot was reached, Langat and I had a thinking session, considering the problem together, and together finding aesthetically appropriate solutions. A stone added to, or taken from a partition, a lower rather than a higher wall, a steeper slope to the roof; a door encased in a wooden frame to blend with the walls; a crafty shelf; the proportions of the raised fireplace, in the Venetian tradition, but with cedar pillars; the curve of a sunken bath. Arap Langat and I shared the pleasure of building something which felt good. We went into the Enghelesha forest to search for a long, dead tree to make into a post, for olive wood for our dining table, or for a huge flat stone to fashion into a seat.

When the house was finally built, the *makuti* roof thatched, the furniture put in place, the brass polished, our antique Ethiopian carpets arranged on the red tiled floors and flowering plants placed in copper pots, I felt a great sense of achievement. Yet I soon realized there was something not quite right, something disturbing and indefinable, some creeping noise which filtered down from the palm thatch and to which I dared not give a name. But Langat did.

'*Memsaab*,' he gravely told me one day, pointing a short chubby finger to the roof, '*iko nyoka kwa rufu.*' 'There are worms in the roof.'

I looked at him pleadingly. He went on.

'*Wewe hapana sikia sawti? Hawa nakula pole-pole.*' 'Don't you hear their noise? They are eating slowly, slowly.'

In the silence – or was it the rustle of the hibiscus in the breeze – I thought I heard the chomping of a million minute mouths

eating away at the roof overhead. I touched the wooden frame
made of *poriti* posts cut out of mangrove trees from Lamu. I
shook it, and an infinitesimally fine powder fell, formed, alarm-
ingly, of thousands of round pellets of digested thatch. It was a
horrible thought and I had to tell Paolo.

When he chose, Paolo could be totally detached. Admitting
the presence of worms in our roof would mean taking cumber-
some, messy and, most of all, immediate action to remove them.

'No, I do not think so,' he said, with a studied indifference
that could not fool me. He too had been listening. He shook a
beam unconvincingly and the powder fell, but with no real evidence of actual worms.

'Dust. I think it is some sort of dust from the brittle leaves.'
We both knew there was more to it than this, and Langat know-
ingly shook his head. But for a time we left it at that.

Yet in the pauses of conversation, when the silence of the
embers followed the jolly crackling of the fire, we thought we
heard the haunting noise of countless creatures chewing away at
our house.

Before long we had to face it.

'What's this?' asked Jasper Evans phlegmatically one afternoon
when he had dropped in for a drink, peering amused into his
beer mug. We looked. A fat pale worm swam in it, frantically.

When another worm plopped in my mother's soup that same
night, Paolo surrendered to the evidence that our roof was
infested by them, and decided to spray it.

Assisted by Langat and his team of *fundis*, perched on the yellow tractor especially driven from Enghelesha, Paolo, like an
ancient squire in a modern tournament, mounted on an improbable *destriero*, manoeuvred the hose-pipe, aiming the potent jet
of insecticide towards the roof.

Inside, all the furniture had been removed and the house was
once again bare.

The exercise took a few days, but the stench lingered for weeks.

At night we ate where we slept, in our bedroom, and during the day under the large yellow fever tree on the lawn. The roof never recovered its original well-combed appearance, and the children laughed, comparing it to an unkempt, windblown hairstyle.

The years went by, and destiny struck twice. On both occasions Langat was in the group of men who came to fetch the grave stones from the bush. Strange to stand straight, alone, watching the crowd of ranch people – twenty? thirty? -- trying to lift them from the place where they had lain since the beginning of the earth. Two boulders – two among millions – chosen to leave the parched bush to mark forever the heads of my men in a garden of green and flowers. Langat oversaw these operations, a small erect figure with a greying round head, and swishing pendulous lobes swinging at its sides.

Since those times, he developed the habit of taking my hand in his for long moments whenever we met, holding it there without shaking it. His slate grey, shining eyes did not smile, but he managed to infuse into them a depth of liking for me which gave me a warm pleasure, as if they were dark friendly ponds, mirroring my smile.

There was a fine dignity and poise about Langat, a quiet certainty of his worth. He looked old, wise, and it was with great surprise that we learnt one day he had taken a new wife. She was a fat girl, with small strange slanting eyes, almost oriental, young enough to be twice his daughter. She spent most of the time standing around the office block at Centre, chatting away with the other wives, knitting complicated patterns in bright yarns, looking well pregnant. Langat went about pleased with himself. His old wife, as is the custom, went back to look after his *shamba* in Nandi.

In recent years, I decided to add to the original hut overlooking the Mukutan stream, now worn by weather and termites, to make it into a permanent retreat for me, a place where I could go and rest alone, and find again my inner voice.

Langat put all his heart into this task: a free house with no doors or windows, open to winds and suns and moons. Perched on a cliff, like a nest, it followed the contour of the landscape, watching over the hills and the green bush, soon to be black with buffalo. It looked down on the stream, beloved by hammerkops and woolly-necked storks, and listened to the frogs and giant toads and the million crickets. Together, Langat and I built in it a sunken bath patterned with shells, a four-poster bed of dead trees rising from the rocks, and a roof to host wandering swallows.

Then one day, when the construction was about to be finished, Langat was taken sick, of malaria they thought, and he was sent to the hospital in Ol Kalau. Before he left he wanted a photograph, of himself, Lwokwolognei and me, within the enclosure of the stone bed-to-be.

A few days later I flew up to Laikipia from Nairobi in the middle of the week with an Italian friend, who had come on a brief visit. I took him to the Mukutan, to have a look at the progress of our work. Nobody spoke when we got out of the car. They lifted their heads, and lowered them again, without a word.

Lwokwolognei was plastering a stone seat, with slow careless strokes. He looked up, his one eye full of pain, like a wounded antelope. I saw with a sense of shock that something sad, terribly sad, had happened. I put my hand firmly on his shoulder, so that he had to turn his long thin neck and look up at me with his one good eye, the empty socket gaping pitifully, as if he were lost, or angry at some evil and incomprehensible god's wrongdoing.

'*Kitu gani? Kitu gani naharibu roho yako?*' 'What is it? What has happened to spoil your heart?' I inquired quietly.

They had all stopped working and waited, and that bated silence that we grow to understand, living in Africa. Only then did I realize something was missing: Langat was not back. And even before Lwokwolognei spoke I knew his answer.

Langat's malaria had been a minor stroke which had been followed by a terminal one. I would never see him again.

'Langat is dead. His hour has struck,' Lwokwolognei said simply, and he went back to work.

I felt in my own heart that familiar squeeze, anger, sense of loss. Another tie with my past gone forever. Another friend passed on. Those wise twinkling eyes, those stretched lobes, his face when Ema had died. Yet what he had built will be there for as long as there are hills.

'*Ni shauri ya Mungu.*' 'It is the will of God,' – the African explanation of the unexplainable and of the unavoidable – I murmured, small and vulnerable in the presence of infinity and I felt, again, the wisdom and comfort of believing it.

'*Ni shauri ya Mungu,*' Lwokwolognei repeated.

They all nodded, in sad patience. Thus, in Africa, one accepts the strokes of fate. I shook all their hands in silence and drove off. My Italian friend had understood nothing.

A Bed Like a Vessel

Give me the life I love
Bed in the bush with stars to see.

ROBERT LOUIS STEVENSON,
Songs of Travel, I. 'The Vagabond'

E ven if, in the West, we no longer normally believe in the truth of auguries, they are part of Africa and of its traditions. I have chosen Africa, and have grown to accept and to respect its rituals and beliefs, as they are rooted in the very nature of its people and in their simple lives, still close to the source of all things. They are, in their essence, indistinguishable from the instincts which allow tribes to survive in harsh conditions, or migrate periodically to other grazing lands, and which protect them from predators, or guide them to water.

There have been occasions in my life when I have been especially close to the depth of the African spirit, and have felt, with humbleness and pride, that Africa has accepted and, in its inscrutable way, has chosen me too.

Like the time when the Pokot women came to offer me a special wish.

It was the afternoon before Christmas Eve in 1983, when Paolo had been dead three years, Emanuele had followed him a few months before, sent to the country beyond by a snake which could not know what it had done; Sveva, our baby, was about three years old. I was in the kitchen, preparing some complicated chocolate log with my cook, Simon.

Rachel, the Nandi maid, had come to call me: '*Kuja! wanawake ya Pokot iko hapa. Unataka kuona wewe.*' 'Come! The Pokot women are here. They want to see you.'

I wiped my hands on a cloth, took my baby on my hip and went out, licking the chocolate from my fingers.

I could see them all from the verandah, some already squatting, at ease below the fever trees, some standing, some dancing round like long-limbed ostriches lifting their legs high. There were old women, covered in soft skins, their hair in dark greased ringlets, toothless, their craggy lined faces worn like ancient wooden masks. There were the girls, and it was they who sang. They were so young, so frail, like birds who have not grown feathers: stick-like legs, thin arms, gleaming with brass bracelets, and their little round faces on pert necks, plastered startlingly with a mixture of white ashes and chalk in ugly patterns. But their merry eyes, glinting with mischief and teasing anticipation, denied the very purpose of their disguise. They each held a long ceremonial stick which they had gone to cut for themselves from some special shrub in the forest; they oscillated them now in rhythm with their voices.

They sang; and when they saw me, their song gained strength, momentum, as if a sudden wind had given it new wings. It was a shrill, high-pitched lament which sounded like the call of a bird at noon. It ran through them like a shiver as they sang in turn, rippling them into a frenzied, yet curiously composed ritual dance.

It was the song of the girls who have been circumcised, and have borne the pain and the ordeal of this barbaric but accepted tradition with courage and dignity, knowing that they were now free to enter in their role of mature women, and to allow a man to find them, and pay their bride-price of cattle and goats to their fathers. It was the song of the women of Africa, a song of courage and mutual solidarity, of hope of children and proud resignation to an unchosen, yet time-proven fate.

They spat on their hands before shaking them with me one by one. They giggled, shyly. Some seemed so pitifully small, still almost children, the white of their eyes and teeth standing out in their tribal camouflage, intended to make them repellent to men for the duration of their recovery. In a few days, healed and ready to be seen by all eager young males, they would wear their traditional costumes of bright orange, brown and yellow beads. Their round cheeks would be greased provocatively in red ochre, their hair dressed in complicated tresses, their little breasts bare. Their skirts of beaded calfskin long in the back and gathered in front would reveal their agile legs, tinkling with anklets.

From now on, in the years to come, I would recognize some of them, surprised at the turn of a track, while they tended their goats and weaners, their stomachs bulging with their first and second and subsequent children, year after year, for all their fertile lives.

Suddenly that day, when all the greetings had been exchanged, my symbolic offering of tea and sugar had been passed on, and the ceremony seemed to be drawing to its end, one old woman came forward, proffering in her hands a long object. An instant hush descended on the small crowd like a last layer of leaves after a sudden storm subsides.

All the other women gathered and surrounded me in suspended silence, their faces alert in anticipation. I understood that this was really the major purpose of their visit, and the core of this ceremony. Murmuring guttural words in Pokot, the old woman offered me her gift. It was a wide belt of soft skin, smeared with greased ochre and goat fat, and beaded in a simple pattern with small grey money-cowries like shiny pebbles. Before I knew it, with endearing giggles, they fastened it around my waist, and asked me something I could not understand, with a happy, pressing, even demanding note in their voices.

'Nyumba yako ya kulala. Hawa nataka kubeba wewe kwa kitanda yako.' 'Your bedroom. They want to carry you to your bed,' Simon interpreted for me, appearing at my side like a protecting shadow.

My bed. I remembered the day when I had sat on the floor of the room which would be our bedroom, watching Langat and his assistant Nguare carrying in the trees that would become our bed.

Building a bed is like building a vessel to carry us through the delights and nightmares, surprises and follies, pauses to rest and deliriums of fever during the varied travels of our nocturnal life. Our bed is the most important piece of furniture we may ever use, for so much happens in it to influence our waking hours. A bed is a *habitaculum*, a home in itself. Our sleeping body entrusted to its protection gathers with abandon the strength, the glory of tomorrow's daylight.

Our bed was a four-poster that I had imagined and built of simple and irregular but polished posts linked by smaller beams and with a sensational head and foot made from offcuts shorn of their bark, to reveal the sinuous curves of their long woody muscles. This bed, where I still sleep and will for as long as I live, was shaped in a few days inside the room. Tall, massive and unique, it can never be moved out through the small door.

We built it together, Langat, Nguare and I, and when it was ready, I gave them a beer, to celebrate that receptacle of dreams and sorrows. We stood together, admiringly contemplating the result of our work, and turned to each other to shake hands on it, grinning with the pleasure of recognizing our achievement. The trails that insects had carved laboriously in the surface of the wood created a subtle, inimitable filigree, a delicate fossil memory.

One day – I was then yet to know – Paolo would climb onto the bed and hang from its central beam an empty ostrich egg, to puzzle my soul with the unrevealed oracle he had concealed inside it. I would spend sleepless nights watching that egg after Paolo died, while his baby daughter would nurse and play on the hyrax cover. A few years later, I would lay there, for his last night on earth, the tortured teenage body of my son, Emanuele, and I would spend the wake, curled up on my side of the bed, rolled into cold blankets, writing for him my last song.

Show the Pokot women to my bed? The request was so odd and unexpected that I had no time to demur. With a nod, I indicated the way, and amongst cries of triumph, I was suddenly lifted high, above their heads in ringlets, by dozens of strong, skinny hands grasping me tenaciously through my khaki clothes, while a new song was being sung.

They carried me, snaking their way in circles through the garden in a live brown stream, like a procession of harvester ants carrying a large white insect to their secret pantries; before I knew it, my room was full of them, swarming at once and all together.

I was finally thrown, as gently as possible but still roughly enough, onto my bed, amidst howls and giggles from the youngest women, while the older ones proclaimed in a half singsong their prophecy, or their wish. One by one, they all fell silent, until only the oldest one spoke. Hovering over me, searching my eyes with her shining gaze, in guttural bursts she pronounced her sentence. They all chorused the last word, clapped their hands and, one after the other, spat on me their blessing in convulsive sprays of fine spittle. Then I was brought up and out again, dazzled, smeared with ochre, to see the sun.

They explained to me eventually that theirs was a special wish for me, that I should again be happy on that bed which had seen such sorrow – happy, and loved, 'but no more pregnant'; the best wish they had to offer, which I accepted gratefully.

And some while after, like all authentic spells, it came true.

Looking for Sandy

In memory of Sandy Field

Great things are done when men and mountains meet.
WILLIAM BLAKE, *Gnomic Verses, i.*

The morning of the 14th of June 1996, dark clouds gathered in the sky over Nairobi, brought by strange and invisible winds. It had been one of Kenya's weirdest rainy seasons in people's memory.

The long rains of April had been delayed, and when finally the sky darkened one afternoon and large warm tears of water fell in untidy splashes, faster and faster on the parched red soil of my garden at Gigiri, we were well into May, in fact into June, almost.

In Laikipia the cattle had grown thin, and the Centre Dam was so low that, with a few shreds of meat, we could gather a bucketful of crayfish in a few minutes just from the shore.

The buffalo I met on the road at night stood dazzled in the car headlights. Rooted on stocky legs, with bent heads, the heavy horns weighing them down, the old males' muzzles lifted towards the car. Opaque ribs showed below dusty hides, a weariness in their eyes, and lions grew alert, I knew, waiting for their chance.

Through the dry sparse silver grasses, windswept in the evenings on the ridge of Mugongo ya Ngurue, I could see Lake Baringo shrink, the banks emerging, brown on the jagged shores; the middle island grew larger, accentuating its sleeping dinosaur shape; hiding in the muddy shallows where fish had long disappeared, hungry crocodiles began to attack goats, and the tree frogs were thirsty and silent in the chilly nights.

In Enghelesha, the fields that Shahar, our young agriculturalist from Israel, had prepared for planting were waiting too; clouds came and went, bypassing us, and the sound of thunder filled the sky, grumbling only from the west over the Cherengani hills. The eastern sky, from where the rains come to Laikipia, remained clear and sparkling blue like wild convolvulus.

The lawn in my garden at Kuti dried partially, and Paolo's dam was empty, the creamy chocolate clay in the bottom cracking slowly in the sun.

The elephant waited for the night to settle before approaching my garden, tempted by the still green shrubs and fresh leaves on the irrigated parts; and the *askari* had a hard time restraining them from getting too close to the graves. Their dung was found practically every morning, still steaming, all along my drive, and as close as Sveva's old playroom. Often buffalo and zebra droppings began to appear too, closer and closer on the murram paths, and my dogs howled their disdain at the renewed invasion of their territory, but their barks were lost in the moonless nights with the cries of plovers.

The concert of birds, converging from miles around to the bird baths and the pawpaw that we laid out on the stones for them, took over any other sound in the scorching afternoons. Hundreds, thousands of birds, covering the dogs' food bowls with an incessant flutter of feathers.

Then, from the garden taps came a gasping sound, and we knew that the water had dried out in the tanks, and we had no choice but to pump it all the way from Ol ari Nyiro springs, well over six miles away.

There was no grazing left; only below the *lelechwa* did some grass sprout still, grey with age and exhausted of life. Even my rain stick did not seem powerful enough this time to stir the clouds.

Finally, brought to despair, I decided to follow the ancient custom of the land, and hired some old *wazee* from Mutaro to perform a rain ceremony. A sheep was found, completely black, with no white markings whatsoever, and, led by Garisha, a cattle-man from Meru whom we had employed on the ranch for years, the old men sang and danced below an old *mugumu* tree next to the Mukutan stream, slaughtered the sheep ceremonially, and proceeded to consume in secret, ritually but with satisfied glee, certain chosen parts.

We waited. We waited hoping to hear the sound of rain on the roof and the frogs waking up in the pond. But nothing seemed to happen.

About a week had gone by when, although no apparent change was detected in the air by us humans, bees started swarming, passing above the yellow fever trees in buzzing clouds which cast fast shadows. On the newly potted palms, on the papyrus in the fish pond, the weaver birds landed to tear off long shreds and, in a frenzy of activity, they began nesting. Soon, heavy bunches like intricate baskets festooned the acacias' branches on the west side of the wind, and my cook, Simon, took to looking east, with straight head and wide nostrils sniffing the air, and commenting gravely: '*Mimi nasikia arufu ya mvua.*' 'I can smell the rain.'

Then I woke up one morning, and knew that the wind had stopped. The air had grown moist, pregnant with water, and the heavy western sky purple and black. Clouds hung, lead coloured, gathering above like a moving blanket.

Thunder and lightning shook the sky, a sudden breeze lifted, the bougainvillaea trembled, livid in the strange grey light, and the rains came.

In Laikipia the termites woke up from their secret tunnels, the earth opened, and golden clouds of living insects drifted high in the last twilight of dusk, in a nuptial frenzy of new wings, dancing dizzily upward.

They did not seem to stop. The dams filled and overflowed;

the Mukutan stream became a river, hurling down the *luggas* and the gorge with a continuous roar. The Demu ya Schule had gone over the road, the gap filled with murky water widened, and the Wilderness School woke up as an island on a livid morning of more incumbent water. The staff, the following night, all standing in their nightdresses on the tables and desks, had to be rescued urgently because the level kept mounting and no one could swim. The land cruiser which went over to fetch them never got out of the water, and we had to send a tractor next day to tow it. Cars got stuck continually in the black cotton soil, and in the gluey red clay alike, and everywhere was flooded.

Crayfish became impossible to catch, now that the Centre Dam had become a mighty lake, refilled constantly by the Big Dam, and, after subsequent showers formed countless running streams, a constant flow of water ran from the overflow, to concentrate into a deafening, rumbling grey river billowing through the valley of Maji ya Nyoka. There is no mid-way in Africa: it is either drought or floods, life or death, it is the land of opposites.

All our rain-catchments were full as they had not been in years. Our sixty-two dams were unrecognizable.

The Red Dam, the Black Dam, Nagiri's Dam and Ngobithu's Dam were fuller than they had ever been. And Mutamayo Dam, Nandi Dam, Enghelesha Dam, Corner Dam, Luoniek Dam, Ol Morani Dam, Kudu Dam and Dam ya Faru had flooded over the road; new islands had emerged everywhere, water ran into the *lelechwa*, waterbucks stood confused on the flooded shores. Even water holes were enormous.

In the kitchen at Kuti a few inches of muddied water covered the floor, where Simon waded ankle-deep and barefooted, rolling croissants impassively, with great style.

In Nairobi, it became a problem to take off and to land in the dense fogs; and often the airfields at Wilson were closed to any traffic for hours.

It was a morning like this, the one of June 14th. I had taken off at 12 p.m. with Benjamin Woodley, our pilot and game warden, who had come to fetch me with our plane.

The clouds hung low; a grey impenetrable curtain covered the Aberdares. We circled round the clouds, going too far south in the process, well off our north-west route. We flew very low, with no shadow, over herds of goats, which streamed away like brown and white seeds scattered on the flat savannah by a hurried hand. Maasai herdsmen clad in red blankets looked up, leaning on their spears, so close I could see the colour of their beads and the knots of their *rungus*, and when, beside the cattle, I began to see giraffe and zebra, I knew that we had gone well below the Longonot crater. We circled and circled, first lower, then higher, trying to find an opening in the barrier ahead. But the sky looked ominous, closed to our probing.

We climbed finally above the clouds, so high that my fingertips became dizzy and my head too light, and I found it hard to keep my eyes open in the blinding white reflection of spun clouds and the sun, too close and too aloof to care for us daring insects. I felt unimportant and vulnerable, suspended in a man-made flying machine inadequate to compete with such powers.

Yet we managed to make it. Benjamin, skilled and cool-headed, finally located a hole in the canopy and dived through the sky as a fish into the deepest ocean; found a gap through the cliffs of Hell's Gate, and we flew very low over the siphons of hot smoking sulphur. Some large buffalo crouching on the grass looked up at us; waterbuck and eland and giraffe and zebra galloped off, but there before us Lake Naivasha glimmered, pale yellow reflections rimmed the clouds announcing tentative sunshine ahead, and we aimed towards it, past Ol Bolossad, up the escarpment.

Far on the right we left the dark Wanjohi area of deep forest and bamboo and igenia, black in a storm of rain.

I looked ahead at the familiar hills on the Great Rift. Enghelesha, Kutwa, Mugongo ya Ngurue. It was the Laikipia plateau, and with immense relief I knew we were safe.

At the same time our friend Sandy Field, piloting his Cessna 172, was disappearing to his destiny somewhere, perhaps in the bamboo and cedar forest, perhaps crashed over the waterfalls, where he may never be found again.

Sandy Field was a man of adventure and charisma, had been a gallant soldier and competent administrator. As a young man he was posted to Sudan, and spoke many local languages fluently. Brave in the bush, where he shot big game, in control and at ease at social gatherings, he had a flair for words, an amazing memory and an extensive repertoire of quotations and poetry. Old-fashioned, cultured, polished, slim and white-haired, with a witty smile, he was a great old gentleman, entertaining and fun to be with.

Now in his seventies, he still flew his small plane everywhere, and was regarded as a cautious, competent pilot.

On the ranch in Laikipia, without a telephone, and having neglected for a few days to listen to the sked on the Laikipia Security radio network, I did not know what had happened. I heard nothing until I flew back to Nairobi a few days later; there, at Wilson Airport in the early morning, I saw a group of Highlands friends leaning against their plane with steaming mugs in their hands, unshaven, silent, wearily eating sandwiches out of a box, sitting on one wing next to some open maps.

'Having a picnic breakfast? What's the occasion?'

The question hung in the air, and then Will answered.

'We are looking for Sandy.'

'Sandy?'

'He flew out of Wilson last Friday for Nanyuki. He never landed. Everyone who has a plane is looking. We have subdivided the area on a grid system. It is going to be tough, but we are trying to find him.'

But it was more difficult than we had all anticipated.

In any other country perhaps, finding a person missing in a storm with his airplane would have been a problem left to the authorities. But not in Kenya. Here the solidarity and friendship that unites the people of the Highlands once again prevailed, and everyone who had a plane or could hire one went looking for Sandy.

I felt I too had to do my share. I liked Sandy. I admired his intelligence, his agile mind, his wit. I respected his courage and leadership. Like his friend Wilfred Thesiger, he was one of the few remaining old soldiers of bygone days, with the soul of a poet, the manners of a gentleman. Kind, eloquent, witty, a strong personality, an addition to every party, quality and humour exuded from Sandy. His disappearance left us all bewildered.

We searched for days and then weeks. Whenever we flew, even after everyone had given up hope of finding him, I kept looking down the secret valleys, the steep walls of rock, the thickets of bamboo and cedar in the hope of finding a sign. But in vain.

All the wildest stories, the most adventurous guesses soared and subsided after a while. Sandy had flown south to Tanzania; had disappeared in the Naivasha Lake; had crashed on Mount Kenya; was alive, in Ethiopia. The last word on the radio before silence had been 'Abyssinia'. We would find him alive up Mount Meru, camped below the wing of his plane . . .

During the search, an old propeller airplane, of whose remote disappearance there was no record, was discovered on the bottom of Lake Naivasha; countless illegal charcoal kilns had been reported, and logging of the rare, precious igenia in unvisited forests, but no trace of Sandy.

Still, we looked from the plane, scanning every crag of the mountains. And then a medium from Europe declared that the plane had gone down in a certain gorge of the Aberdares next to the Wanjoi valley, and, to clear our conscience, we decided to go and look.

We left at sunrise from Laikipia, Aidan and I, on a dark morning of late June, and drove, through muddy tracks, up green hills and mountains. At the park gate we took with us a ranger who knew the way, a young Tugen from our boundary of Enghelesha, it transpired, trim and fit, well equipped with rucksack and tentage, and a very good uniform. We got stuck a few times up hill, walked in the red mud to put branches below the wheels and were astonished to see so many tracks of lions and hyena spoor. I wondered what they ate, as there seemed to be few signs of buck.

Lions were not indigenous to the Aberdares. They had been recently introduced, and – ecological tragedy – had multiplied beyond balance, almost completely destroying the rare and shy bongo antelope that feeds on the moss growing off fallen trees in the cold and rainy mountain slopes. In recent times, drawn by hunger, they had been known to attack and kill people, and I knew a scheme to control their numbers was under way.

Finally, we reached a fog-covered peak, where for short seconds when there was a break in the clouds, a sensational landscape intermittently appeared below us. We set camp at dusk, in the middle of the track, as the sides were thick with impenetrable vegetation. There was a drizzle of rain which soon grew to a mighty downpour; every inch of tent was wet and the mattress was swimming in chilled water. We ate a scalding bowl of soup, and piping-hot stew to revive us, cooked on a small camping gasstove that I had providently brought, for our campfire would not burn with wet wood, and, after a hot, sweet rum punch, we slept surprisingly well.

In the morning the search began down the crags of the land, over the slopes and inside the bamboo thickets. But it was in vain, and at night we finally returned, convinced that Sandy would never be found.

The bamboo, we knew, would have closed, resilient, over the plane and would have grown many feet in few days, obliterating it from view forever.

Everyone knew that the forest was full of wild animals, and hyena are unkind to the wounded and weak. We hoped for a quick end, a peaceful rest in the unknown forest grave.

In *absentia cadaveris,* a memorial service was celebrated for Sandy in the church in Nanyuki, resplendent with elaborate flower arrangements by the accomplished ladies of the Gardening Club. There was all the crowd one sees at weddings and at funerals, at the polo matches of the North Kenya Polo Club, and at the Beating of the Retreat in Nanyuki, when some kilted Scottish battalion marches back and forth in the drizzle, with impeccable martial choreography, to the lament of Highland pipes. Everyone knew everyone else, all men wore identical blue blazers over ill-fitting trousers, the women assorted frocks and flowery hats. There were old settlers, farmers and retired game wardens, old-timers, and, finally visible, the faceless voices of the Laikipia Security Network.

Sandy's friends delivered moving eulogies, like Tony Dyer, a fellow pilot and a contemporary, and few eyes were dry. Everyone sadly agreed that we would never know where Sandy lay.

But we were wrong.

The honey gatherers found him.

They had been up in the forest slopes, smoking out bees from the old olive trees across the clumps of bamboo. They heard thunder meeting earth, and the shudder of the impact. They ran towards it, warily at first, and then, spying through the dripping leaves, they saw it, the white bird of metal that had fallen from the clouds and crashed miserably, a wounded seagull, sinking through the tall green foliage like a boat at sea.

They watched breathless for a while to see what happened, what creature would erupt from the contorted wreck. There was only silence, except for the shrill calls of invisible monkeys up on the treetops, the sound of water rushing from the bamboo.

In the impact, Sandy's life had lifted from his body and joined

what was before, and after. Over the purple-ink clouds of the
Great Rift, gliding along the waterfall's drops covered in moss,
over the silent lakes bejewelled with flamingos, over the valleys
and savannah, riding on the long necks of giraffes galloping in
slow motion on the plains, or amongst trumpeting elephants in
the dust, up on the mountain slopes and up to the dazzling light
of the equatorial sun, like a butterfly finally hatching from its
cocoon, what had been Sandy's energy, now freed from his body,
exploded to blend with the Whole.

The men approached tentatively, climbed over the wing and
spied inside: the *musungu* reclined on the steering wheel, his
white hair visible, his forehead resting on the dashboard splintered in a thousand diamonds of exploded glass. He looked
asleep, but the two men knew with the wisdom of the bushmen,
from the shape of the neck, that the white man would never move
again.

There was no harm, they thought, to find what he had no
longer use for – clothes, watch, shoes, money perhaps? – that they
could well use. It had been a bad year; life was hard on the mountain. The last honey harvest had been meagre and it looked as if
the next one would be too. What if they helped themselves to
what would rot anyway up the mountain?

No one would suffer and no one would find out. They cut the
seat belt and dragged him out of the plane. He was light. Gently
– there was no hurry – they put him on his side, and searched his
clothes. They worked quickly, without talking, taking everything
they could. His briefcase, all his clothes, his shoes. Then they
covered the wreck of the plane with branches to hide it from the
sky, for even in their ignorance they knew there would be a
search. The more delayed the finding, the more chances they had
of not being found. The forest animals would come. It would be
easy, quick. Hyena were always pacing the game trails at night
with their loping gait, sniffing for food. The forest would swallow

the *musungu* man as it was now swallowing them, thick, impenetrable, keeper of secrets.

Loaded, happy, they ran off, the honey search forgotten for today, to divide the loot in a safer place.

But it was not the hyena.

When the night came with its cloak of fog and drizzle up on the mountain slopes, and the night frogs woke up in a choir of life, the old leopard came out of its den under the boulder of rock, hung with fern and wild figs.

The leopard sniffed the air and moved slowly and surely through the undergrowth, towards the strange white shape in the bamboo.

And it was the leopard who carried Sandy off to the tall tree.

The forest men, months later, having learned of the reward offered for Sandy's discovery, went to report to the nearest forest station. A group of friends in the helicopter lent by Halvor Astrup went to land immediately on that spot, which no plane or car could have reached. Traces of hair on the dashboard and the cut belt told their story, and so did the leopard spoor abundant round the wreck.

Now Sandy lives in a leopard up on the mountain slopes, a healthy strong lustrous leopard, and we know that he would have loved that.

Aidan's Return

I am waiting for a welcome sound, the tinkling of his camel bell.

ISOBEL BURTON, Letter to Lady Paget

The magic of the Highlands of East Africa is the evenings, when the noises and the colours and the very essence of the air change, and a wind begins to blow in sudden blasts, bringing with it tales of places far away.

It is then that we can imagine anything may happen, and while the sun takes on a light of deep red before setting, all the memories, all the prayers, and all the tears we have ever shed flock around us, squeezing our hearts with often unbearable pain.

I walked out facing the wind, all my dogs running ahead, in a flurry of tails and joyous barking. I walked up the airstrip at Kuti towards the green hills where euphorbias grow thick, and watched a herd of elephant slowly moving through the bush, towards the water-tank hidden by trees.

The dogs took off, all at once, after a gigantic male warthog, their fierce barking swallowed by the growing shadows. I found an old anthill, big enough to sit upon, and crouched there, gathering my shawl about me, my feet on a twisted *mutamayo* bleached by age, to write my diary. Harvester ants hurried down their hole, carrying the last yellow seeds of the day. In the crepuscular sky, huge clouds shone, of the deepest coral. The hill of Mugongo ya Ngurue looked black in the twilight.

So often, over the years, this had been my familiar evening walk. Alone; sometimes with Sveva; in the recent past with Robin, and his hair had been the colour of the bleached grass at sunrise, his laughter fresh, sincere and a balm on my wounds. Since our paths had gently parted as it was written, the dogs were my only companions, and my deep longing for Aidan, the man who had eluded me.

A couple of years after Paolo's death, and a secluded, lonely life filled only by the presence of my children – Emanuele, then a teenager, and Sveva, who had been born after Paolo's fatal accident – destiny had one day guided me to a country wedding, and there I had met him for the first time.

Tall in the crowd stood my love, and I recognized him. In his strange, secret way he recognized me too. For a time we shared the music of my sheltered room, we knew our faces by candle-light, and the smells were of incense, of love, and cut flowers in vases.

What I knew more about him was the intent look of his stirring blue eyes, the deep voice in the poems he read to me, the feeling of his strong, demanding body, his arm outstretched from the window of his white car in the grey light of early morning.

I gave him all I could, but the time was not ripe yet. He had to go, leaving me waiting, at the bottom of my solitude, but taking with him the key to my door, just as he held tight, already and forever, in his long fine hands, the key to my inner self.

Of Aidan I missed the masculine presence, the deep voice and the searching eyes, the subtle poetry and inexplicable appeal of the solitary nomad, the adventurous man who walked alone and to whom the wild was familiar. Aidan had no fear in the bush. His confidence came from his knowledge and love of things untamed, of plants unknown and of paths not yet trodden by human feet.

I missed him with a longing to which I could give no name. With a patience which was most uncharacteristic of me, and with unreasonable faith, I waited for years for the time when he would be back. Evening after evening, when I was in Laikipia, I walked up this airstrip with only one hope. If the power of my yearning could be a magnet, I knew that he would be drawn back. When the time was right, I would be ready.

I listened to the noises that every African night always brings: guinea-fowl and nightjars, a far whine of hyena, a faint mooing of cattle being herded to an invisible *boma* behind Kuti hill. The elephant, the largest, were as ever the quietest of all. Only a broken branch and a stomach rumble betrayed their closeness. On the horizon a white sliver of moon began to rise.

I waited quietly, with a sense of encompassing peace, for the friendly darkness. My dogs, now back, formed a protecting circle of warm panting bodies round me. Silence. From the top of the termite hill, I could watch with no fear.

The sound came suddenly from behind the treetops, in a still pearly sky. It was like the distant buzz of a persistent insect, approaching fast and growing louder in the stillness of dusk. I knew instantly what it was and at the same time I saw it. A small white aeroplane approached from the east, suspended in the sky, flying low over the trees, gliding over the hills, and straight towards me.

It was too late for a small plane to fly. In a few minutes it would be dark; nowhere could a small plane land over the shady precipices and valleys of the Great Rift, unless . . .

I stood slowly, and all the dogs with me.

The wind, again, took away the noise and there it was, circling over Kuti hill, pointing towards Nagirir, lower, much lower, white wings outstretched like a bird flying home. Before I could gather my thoughts, and calm the surging emotions and the wild thumping of my throat, here it was, landing in a cloud of dust.

I walked uncertain to the centre of the strip, my back to the silver moon. In the last glimmer of dusk the plane glinted, turned towards me and came to a stop. Shading my eyes, my heart just pounding, I moved slowly towards it, unbelieving. I had dreamed for years that this might one day happen.

A few weeks before, there had been a letter, inserted in a musty old book, a rare first edition of an autobiographical novel by his favourite uncle. So enthralling was the story, and so masterly the passionate style of writing, that the book had been haunting me since, as had the note – the first in years – with its elusive promise.

'I often talk to you, who are sitting on my shoulder. Things have changed. One of these evenings I will come to find you. If you would ever . . .'

Now there he was, as ever unexpected, landing for the first time on the strip I had built for him in past days of misery, as only happens in those tales in books.

Even before the tall shadow jumped out, I was running. I stopped a few steps away. He had changed little: a slender living statue with broad shoulders, searching eyes in the serious sun-burnt face, straight nose, tight curly hair, firm soft lips. He moved a step. I moved a step. We moved together, and he caught me, crushing me against his breast without a word.

'I'll never leave you again,' he whispered on my mouth.

So came back Aidan.

The Magic Cove

And hand in hand, on the edge of the sand
They danced by the light of the moon.

EDWARD LEAR,
'The Owl and the Pussy-Cat'

The breeze from the ocean moved the palm leaves, and the hair on his high pure forehead. Straight and slim in the ceremonial starched white uniform of the Royal Navy stood Charlie on the beach at Takaungu, on the afternoon of his wedding.

I had arrived breathless after a long flight from Nairobi with Sveva, a short pause for lunch in Watamu, and an unending sweaty queue on the jetty at Kilifi, plagued by cashew-nut peddlers.

The smell of seaweed, the steamy heat which lifted from the creek, the colourful crowd waiting for the ferry-boat – fishermen, children and the bare-breasted, wide-hipped Giriama women wrapped in bright *kangas,* balancing baskets on their heads – were characteristic of Kilifi. The strong aromas of mango, dried fish, coconut oil, ripe banana, perspiration, smoke, jasmine and sandalwood blended in a heady scent. I inhaled deeply; despite the heat, I always enjoyed being there.

The afternoon had gone by fast, almost without us noticing. In the end, short of time, and continually overtaken by cars loaded with people dressed for a party and clearly going to the same wedding, we had chosen to change in the bush, below a clump of palm trees, amidst much giggling. We stopped the small hired car at the side of the road, so that Sveva could slide into her beautiful dress of cream satin and pink summer roses.

It was hot that day of December, and sticky as could be. The guests had already taken their place down on the beach, in the sheltered corner of coral rocks and fine white sand that Charlie and Emanuele had long ago called 'the secret garden'. I surveyed the audience.

People sat on neatly arranged rows of wooden planks set on white stones. A barefoot Mirella stood on one side, clad in purple and crowned with a garland of frangipani like an ageing nymph, taking photographs and looking out at the choppy sea. The altar was just two large candles and a handful of shells and white flowers scattered on a large trunk of sea-bleached driftwood. In front of it was the couple.

A knot of emotion rose in my throat, as there, next to his lovely bride, stood my Charlie, clothed in so many memories, handsome and lean like a young Mountbatten, romantic in his white uniform with golden epaulettes, the familiar young boy's smile brightening the unchanged charming face below the lock of hair. Emanuele's best friend.

I could well imagine the shadow of an adult Emanuele, as his tall, good-looking best man, standing right behind him.

That evening we went to the wedding party, and the baobab that we had once called bewitched, was silver in the moonlight. The place was the house on the cliff next to the one that we had regarded as ours. It had once belonged to a strange lady who lived alone with many cats and loved plants and growing things, but was devoured by a silent unhappiness, which she tried to tame with pills. One night she could bear it no longer, and she had swallowed them all. The house was then abandoned, for a time, to the sea swallows and the salty winds, and strange tales about it were whispered by the Giriama house-boys, and by the fishermen who came along a low tide to sell their catch of lobsters and coral fish.

Eventually the house was reopened and inhabited again, new tenants lent it their fresh auras, and the shadows receded. Tonight the place was animated, garlands of coloured lights were strung from palm to palm to fight the black of the night. Music drifted high with the sea-breeze. The air, balmy with nocturnal jasmine, carried the voices of the festive crowd, and everyone was merry.

There were people I had not seen for years, the Kilifi crowd who used to come to our parties, the men who used to speak with Paolo of fishing, and even Mohammed, the retired old barman of the Mnarani Club, who for two generations had reigned behind his counter, knew all our children by name and, despite being a Muslim, remembered everyone's favourite drink. They were the times of the Lady Delamere Cup, when the boats assembled at sunrise to depart for the high seas, following a dream of marlin or sailfish, shark or tuna – or, at least, a great barracuda – and came back at late noon with red, blue and yellow flags flying in the wind, which we all tried to interpret through our binoculars to see who had caught most.

Then, Pimm's after wave of Pimm's, came the weighing and the marking on the old school blackboard. The winners were handed out prizes by a regal and cool Diana Delamere. Everyone clapped and talked about it for weeks. They had been other days, now gone forever, I could well see that. Then Diana had died, and an era with her.

The Mnarani had been sold to the tourist industry, and the old charm of days gone had vanished in the anonymous crowd which changed weekly. Paolo had died and, a few years after, Emanuele too. Charlie, the companion of Emanuele's school days, had been at the time still at military school in England. Soon afterwards, like his father before him, he had joined the Royal Navy. He had kept in touch, and when he was in Nairobi his tall lanky figure never failed to appear on our doorstep. I cared for him.

Dressed now in immaculate light white linen, his trim waist

circled by a bottle-green cummerbund embroidered with a
golden dragon, Charlie sat me at his right side, like the mother
he had lost, and, like the son I had lost, I doted on him. His
brown eyes, gentle twinkling eyes, brimmed with tears of remem-
brance and happy days.

'Remember when you told us about the baobab which moved
in the full moon, and we believed you?'

'Remember when we celebrated Iain's birthday at the end of
Ramadan down at the Mnarani, and Oria asked all the village
women to cook ceremonial food and to lend their *buibuis*?'

'Remember when we found that gigantic puff-adder across the
track below the Fielden house, blocking the way, and Emanuele
refused to drive over it because we would have killed it, and we
had to wait until he had coaxed it to move away slowly?'

'Remember when you gave a surprise party for Paolo's birthday and mine in the cove in the Kilifi Plantations? You had the
cove lit with hundreds of candles, and you marked the line of the
high tide with rows of lanterns, and you invited all Kilifi, and
they all came!'

The magic cove. How could I ever forget? I fixed my eyes on
the champagne glass I was holding, and through its golden bubbles, as in a yellow crystal ball, the memories swept back, of
happy days.

When the ocean was green, with white rims to its waves, and the
trade winds blew in Kilifi, Emanuele went sailing with his friend
Charlie. I watched from the shore for their frail craft to pass,
sitting under a giant baobab tree in our garden. This was a vast
tree with a grey trunk spun with silver which seemed to absorb
the heat of the sun like a human body, and it was my favourite
refuge at the coast in the long afternoons.

They lifted their arms when they saw me, and the slimness of
their young figures was emphasized by the swollen sail and the
immensity of the ocean. Their boat cut away fast, bobbing on the

waves in white sprays of foam, and disappeared behind the coral cliffs, leaving only an empty reef, and the colour of my wonder.

'I have discovered a fantastic place, Pep,' Emanuele told me one afternoon on his return, while still drying his damp blond fringe off his forehead. A glint of enthusiasm lit his dark eyes. 'A small cove, off the Kilifi Plantations. Charlie and I think you must come and see it. It would be great for a party.'

Next afternoon, we drove there together. It was not easy to find it from the jagged shore, as the terrain was covered with spiky sisal, taller than a man, and tangles of grassy twine. We located it finally, at the end of an inconspicuous trail. The boys helped me to climb down to it through rough old coral rocks hung with sea grapes, and we were there.

It was a semicircular cove of perfect proportions, surrounded by rugged grottoes at many levels, where one could instantly imagine hiding bright candles.

The tide was coming back, beating the shore of untarnished white sand, and decorating it with rows of seaweeds and coconut shells in lacy patterns. Seagulls flew low, with still wings and high calls piercing the evening. There was a rare, ageless purity about that place which enchanted me, and I could see that it would indeed be ideal for a special celebration.

It was the beginning of December 1979. Charlie's birthday had just been, Emanuele's would be in January, but Paolo's birthday was due in ten days or so, just before Christmas. There and then, we decided to give him a surprise party in the magic cove.

There began days of excitement and secret preparation. I drove to the bazaars at the market in Mombasa, and bought mats of woven palm leaves to scatter on the sand, and bright cotton *kangas* and kapok to be sown into large cushions. From the workshop of the Kilifi Plantations, through which we had access to the cove, we borrowed steel drums and wire mesh for the barbecues, long low tables of driftwood, and planks of sawn timber to use as an improvised ladder to get down to the beach.

Early in the morning, when Paolo was out fishing, or in the late afternoons, we sneaked away to prepare the feast, to which we had invited, sworn to secrecy, everybody in Kilifi. We cleared the beach of debris deposited by generations of waves, swept the rocks of dried weeds and sand and finally started bringing down the assorted paraphernalia that we needed.

Then the day came, and I told Paolo that we had been asked to a special party on the beach. Puzzled and curious, he accepted happily.

We had worked since early morning carrying and decorating. Now the place was transformed, bright with candles, twinkling from the shelves of stones as in a fairy land. Paper lanterns marked the line of the high tide, and mats, covered in cushions of bright blues and turquoise, were strewn on the damp sand in a large circle round a roaring fire. Hurricane lamps and bunches of frangipani hung from stumps of driftwood. A music played with the sound of the wind.

A vast barbecue glowed in the largest cave, where meats were being roasted by my cook, Gathimu. Large platters of pizzas and samosas, oysters and kebabs, garlic breads and cheeses, mangoes, papaya and pineapple were set out on a long table covered in banana leaves. Wine and champagne bottles stood in a wooden basin filled with ice, and a spicy punch of rum was served in half coconuts garnished with red hibiscus from a bowl with floating flowers. Up above, millions of stars glinted in the equatorial sky. Large stars, exotic, yellow, alive like campfires at night.

As hosts, we were the first to arrive. Realizing what the plot was, Paolo exclaimed in surprise and swept me up in a big hug, laughing. The guests started coming in groups. In all their faces we read the same wonder.

It was a happy time, never to be forgotten.

I did not know yet for sure, but perceived it from early signs, that deep inside me, in the secret oyster of my womb, a new life was forming, made of Paolo and of me to remind us forever of our blending.

In the grottoes, amongst orange crabs, curious and tentative, candles brightened the darkness; mermaids sang, hiding through the grey-green waves; nocturnal seagulls called, and in Paolo's eyes I could imagine winds of unanswered questions.

I did not know – how could I, I who had still so much to learn? – that the luminous sadness in the visionary glimmer of his eyes was a premonition of a future which he would not see, and that this was to be his last party ever.

It ended when the lorry did not stop and Paolo's car did not brake in time, and his life was submerged and taken from his body, to fly to join the seagulls he loved, and the sky and the clouds and the hills of Laikipia.

It ended when the night tide came in, flushing away in its gentle waves the red bougainvillaea, and the candles in the lanterns died off one by one, taken by the ocean, until we knew it was time to move on.

Next morning we went to look. Only some smouldering ashes were left, some dying frangipani still perched on the rocks with the seaweed and an empty wine glass rolling quietly on the shore, miraculously intact.

My eyes focused again on the chilled champagne glass on the white tablecloth. Charlie was looking expectantly at me. Across the table, radiant in cream silk and pink roses, Sveva, the living portrait of her father, watched me with her ocean-blue eyes.

I shook my head and came back to the present. I smiled at him:

'The party at the magic cove. Of course I remember. How could I ever forget?'

Later in the evening, when the dancing began, Charlie asked Sveva, as he would have his sister, and they twirled around in a lovely waltz, the child-woman in her glow of blond hair and silks, stars in her eyes, and the tall youngster who could have been my son.

On the beach below, the tide was again receding, and on the sand it left patterns of shells and fishbones, driftwood and tangled seaweeds, fragments of ocean stories to be interpreted.

Like tides we come, we go, leaving the memory of our footprints on the hard shores of life. I watch their tracks in reverence, holding my daughter's hand, and try to evoke for her, with care and tenderness, their frail loved figures, from the mist of time.

Part II

BRIEF ENCOUNTERS

Full-Moon Island

What if this present were the world's last night?
                    JOHN DONNE, *Holy Sonnets*, xiii

W hen we went to live in Africa, we spent our first Christmas
       holiday at the coast. Although the magnificent beaches
and pure coral reef were already beginning to attract the attention
of international tourism, it was early days. The coastline of Kenya
was largely still the kingdom of seagulls and turtles, of wooden
dhows built from hollow tree trunks, of Giriama and Swahili
fishermen singing to the tides a song of waves and hope of fish.

A few local pockets of quiet middle-aged people of European
origin were attached to costal villages. Their favourite spots were
Malindi, Kilifi, Vipingo, Shanzu and Shimoni.

They were an unusual community of retired residents, mostly
former farmers, who had sold their up-country properties, at the
foot of Mount Kenya, in the dry windswept Highlands or in the
green tea and coffee districts of Kericho and Thika. Undisturbed
and undisturbing, comforted by an assortment of dogs and generous sundowners, they now spent their sunset days in the breezy
shade of their spacious verandahs. Their new homes, built mostly
of white-washed coral blocks, with tall roofs of thatched palm
leaves, were immersed in intricate gardens of bougainvillaeas and
mango trees and graced by a constant view of the shimmering
reef.

They owned boats of various descriptions, grand ocean cruisers, or modest homemade catamarans. which they tended meticulously, since they all shared a consuming passion for deep-sea fishing, or sailing, or both.

The sea had always held a peculiar attraction for Paolo, who loved to explore it, and before we found our own promised land we spent much time there. For people who lived alone in the silence of their memories, and who could understandably have been diffident of strangers, the Kenya coast community were sociable, and made us instantly welcome. Perhaps the fact that we were young, enchanted with the ocean as they were, exotic, carefree and in love, with charming, sunny children, and all the time in the world, triggered their curiosity and an unexpressed nostalgia for days gone by. They offered us, with total generosity, the hospitality of their homes, boats and drinks cabinets, and the friendship of their pets.

With unexpected inventiveness, they devised occasional entertainments to interrupt the monotony of their unbroken days. One night in Shimoni, just before the New Year, they asked us, and my mother who had come to visit us in Africa for the first time, to a full-moon picnic dinner out in the ocean, on an island which only emerged with the low tide.

'Shall we go?' Paolo asked, a blue glint in his eyes. 'The night is going to be clear, a fantastic light.'

A teasing note: 'Yet, it is rather far. It might be wet . . . late for the children and your mother. We'll have to use a compass to find the way. It sounds mad.'

It did. And irresistible.

'*Andiamo*,' I said. Italian was all I could then speak.

We assembled at dusk on the shore below one of the houses. Cool boxes and baskets crowded together, packed with food which revealed the origin of their owners: pickled herrings in dill and deadly schnapps for the Scandinavian; smoked salmon sandwiches, scotch eggs, blue cheese and beer for the English;

white cheese, olives and divine ouzo for the few Greek; and for us, of course, pizzas, salame, provolone and flasks of red Chianti and chilled white Fol. We also had the large classic panettone which is the taste of any Italian Christmas, and which my mother, now slightly bewildered by this adventure, had heroically carried all the way from Venice.

Only in groups made up predominantly of British people can a 'quiet excitement' prevail: active, efficient, aiming at a goal. In this atmosphere the boats were loaded, darkness fell, and we went.

The humid warmth of a salty breeze beaded our faces, and the black oily surface of the ocean opened smoothly to our bows, and closed in a wake of glimmering foam. The plankton shone phosphorescent like submerged galaxies, drawing patterns on the disturbed surface. The engines droned on. Emanuele, a little boy of six, came to rest close to me, and the night was reflected in his large dark eyes, which absorbed everything. Someone sang a slow song which blended with the talking voices, and the engine noise, and the smell of seaweed.

For hours we went far into the blackness. Then, at a point in the ocean which seemed identical to any other, the leading boat came to a sudden halt, its engine slowed to a murmur. We all grew silent, and watched in bated expectation while the horizon glowed lighter and the ink of the night became a blue velvet drape on which stars gradually paled like dying candles. As the breeze seemed to grow into a strange wind, and the current lapping the sides of their boat to be sucked away faster and faster, the largest white moon began to rise, sailing over the horizon.

By and by, massive and silent, the mystery island began to appear in front of our eyes. First the coral rocks emerged, like the crested back of a sleeping sea monster; then a startlingly white, smooth beach of opal began to materialize, matching the cold moon. Dinghies were lowered from all the boats, efficiently loaded with the food baskets, barbecues and crates of bottles. People started eagerly rowing ashore.

I was allotted a tiny yellow rubber dinghy, wet and slippery, in which I sat at the oars, with Emanuele, a case of beer and a cluster of bananas. I started rowing, but had not appreciated the speed of the current. The wind blew stronger. Twenty minutes went by, and I did not seem to have made any progress. I was drenched and cold. Our friends seemed far away, the shore unattainable, and the wind did not carry my voice.

Paolo finally came to my rescue, laughing. In no time I was ashore on the firm, cool sand, safe in his arms, drinking a glass of wine.

People were scattered in small groups, divided or united by their age, taste, spirit, hunger or thirst. The charcoal was lit with some difficulty, helped by the blue flame of paraffin, which the wind soon transmuted into a glowing orange warmth on the wrought iron barbecue. The smell of grilling marinated chicken, fizzling sausages and garlic bread came in whiffs of fragrant smoke billowing on the wind.

The chilled oily schnapps, drunk enthusiastically in tiny glasses, raised the spirits high in minutes, chasing away the shivers, while the jolly noise of popping corks filled the night.

A group of children sang with a guitar. Others ran along the beach, pursuing crabs, Emanuele went alone with his torch, looking for imprudent cowries left out on the shore by the receding tide. Even my mother seemed to have found someone with whom to talk.

Mellowed and content, I unrolled a straw mat and sat on it thinking my thoughts, wrapped in a *kanga*, watching Paolo turning the grilled meat with hungry dexterity, opening bottles, chatting away in English, totally integrated.

Hours went by, and with time we grew silent. A subtle muting again, a shudder in the breeze. The tide was coming back. Slowing at first, then faster and faster, the waves began to reclaim the sand, inch by inch. And with the advancing water, changed our mood.

Out on the invisible full-moon island in the middle of the Indian Ocean everything seemed possible. Was the rest of the world still there?

Italy, recently left, now seemed so far.

Mad thoughts of returning to find no Shimoni, no more shores anywhere on which to land, just a limitless ocean where tongues of beach like this one appeared only for a few hours with the moon. Forever vagabonds on the wild seas like the Flying Dutchman.

'*Passa la nave mia con vele nere . . .*'*

Time to go back. A sudden forewarning, a void, a fear.

I looked for Emanuele. He was running after his dreams in the wind along the shore, minute and unreachable as an elf in a fairy tale, and his hair had the colour of the waning moon. With a squeeze in my chest I called him, and my voice came back to me in the night, like a lost seagull's.

Then Paolo was with him. They were running together, and they were holding hands.

*'There goes my vessel with the black sails.'

The Bull Shark of Vuma

The old man knew that the shark was dead, but the shark
would not accept it.

ERNEST HEMINGWAY, *The Old Man and the Sea*

Out in the Indian Ocean, off the north Kenya coast, between
the fiords of Taka-Ungu and Vipingo, are the shallows of
Vuma.

They are banks of flat, submerged prairies, covered in long
seaweeds growing out of old coral gardens. Restless underwater
currents constantly stir the pale green-grey blades of the plants.
They shudder, tossing languid manes, like savannah grasses bent
by the invisible fingers of the Highland wind. Large shoals of fish
of all descriptions come to graze here from the black depth of the
ocean, like the gazelle and antelope on the Enghelesha plains.

So, too, come the predators.

The waters of Vuma are notorious for the abundance of
sharks, which lurk in the darkness around the shallows, emerging
swift and deadly to prey on the foraging herds. Like all carnivores
and scavengers, they are attracted by the jerky, uneven movements which reveal creatures in distress, and by the smell of
blood, carried through water in thick clouds of livid red, like a
scent on a breeze.

Vuma had acquired a sinister reputation when a number of
dhows had sunk in a series of storms, and the shipwrecked crews

had been devoured by sharks. Such stories were invariably told when someone mentioned the place.

Fast-swimming fish, minor predators, like *cole-cole* and tuna, were known to favour Vuma, and large rock cod lived in coral caves along the edges of the submerged highlands. In a curious way, Vuma was a sea equivalent of Laikipia with its plateau perched on the edge of the Great Rift Valley, and it was naturally irresistible to Paolo, who often went out there from Kilifi with his spear gun, on his rubber dinghy.

He sometimes took a friend or two, but mostly he went with Ben.

Ben was a local fisherman of the Swahili tribe and, like most of them, he followed the Muslim religion, and always wore a tiny embroidered cap. The Swahili, with their Arab blood, distinguished themselves from the Giriama, the most common tribe in Kilifi, of purer Bantu descent.

Ben was small but muscular, with large shoulders on his compact body. Above his short black beard and flat, pointed nose with wide nostrils, his tilted eyes were intelligent and mischievous. He radiated a complete faith in the excellence of his seafaring skills. The arrogance of a race which is noble and proud of its traditions, accompanied a certain roguish indolence which one condoned, as it was mellow, like most coastal sins. It went with the rarefied climate of the coast, the ripe smells of vegetation and humid sands, fruity mangoes, coconut milk and frangipani, *korosho* nut roasted in the evening, fish dried in the sun, spices and damp monsoon. Ben was part of Kilifi, and seeing him meant we had arrived.

With the African's uncanny gift of knowing without being told, and of appearing unexpectedly at the right time from nowhere, within less than an hour of our car arriving at Kilifi, Ben's voice could be heard from the kitchen, greeting my Nairobi staff in his sing-song coastal Swahili. Then the shuffle of his naked feet approached and he appeared, hand stretched out in greeting, calling out our names, asking for a cigarette. We liked him.

Often, his eyes were glazed, his manner slower, dreamier, and the aroma of a reefer drooping from his lips disclosed that again Ben was smoking *bhang*. This habit which, he maintained, made his sight sharper and helped him to see fish better under water, was accepted as part of him, and won him the nickname of 'Bhangy-Ben'.

He was an excellent, natural fisherman, with a genuine passion for the ocean. More than anything he enjoyed going out looking for marlin, and he was unbeatable at spotting the shoals of sardines signalled on the shimmering horizon by fluttering flights of feeding seagulls; they were infallibly followed by hungry *bonito* tuna, and these often by sailfish or marlin.

He knew the secrets of the tides, and the habits of fish, rather like the trackers of the Highlands knew the game they were stalking.

In fact, I always associated Ben with Luka, our inimitable Tharaka hunter and Paolo's companion of countless adventures, who knew how to think like the buffalo. He could find them in thick bush just by listening for the tick bird, just, one might have thought, by sniffing the air with his sensitive nostrils.

Both had the same self-assurance which came from the total mastery of their skills and complete understanding of their background, the Indian Ocean and the Highland savannah. Each in his own way believed his presence to be – and possibly it was – indispensable to any of Paolo's expeditions.

Ben had presided over most of Paolo's fishing adventures: the time of the giant rock cod, and the time of the black marlin which Paolo caught without fishing chair or belt, standing for hours in a rubber dinghy oscillating on the ocean's swell, and which won him his first fishing trophy.

The house where we stayed in Kilifi belonged to an Italian

friend who lived in a hacienda in Argentina. He had only been
there once in ten years, while it was being built. With Latin generosity he gave us the unlimited use of it, and we regarded it as our
home at the coast.

The garden was an unruly profusion of bougainvillaea and
solanum, with some palms and the most magnificent baobab, a
giant of perfectly harmonious proportions, with which I felt great
affinity. I gave it a soul, like all trees, and a soul that I liked. I
spent many hours each day sitting with my back to it, thinking
my thoughts, writing my diary, waiting for Paolo and Emanuele
to come back from fishing. It was a time of peace, in which I
managed to catch some intuition and hold some passing verses
before they flew off with the wind.

Like Ulysses in other seas, and all the sailors ever, Paolo and
Ema so often had an adventure to tell. The giant cod harpooned
with Lorenzo Ricciardi; the dolphins who had joined in an ocean
dance; the sailfish which had gone with the bait; the *upupa* that
had flown from nowhere and landed on Paolo's hair.

And then there was the time of the bull shark at Vuma.

Paolo had gone fishing for *cole-cole* at Vuma one morning of
January, with his brother and a friend. It was still the time of his
passion for spear fishing. Without bottles, with only his mask and
snorkel, he dived confidently and deep. I feared always that his
lungs would burst; but he would come up after what to me
seemed ages, not panting at all, with a fish on his spear and a
blue look of triumph on his tanned face.

Ben could not go out that day. He had come in early to
announce that another of his children had been born in the night
to his sweet Swahili wife, another boy; amongst Muslims this was
a serious matter and cause for much ceremony and celebration.
As he could not go, he advised that everyone should stay at home,
or join me in my planned shopping expedition to Mombasa har-
bour.

Ben did not like it when he could not go out fishing with

Paolo, and always made sure everyone knew of his disapproval –
as if his presence were essential for every adventure to be faultless,
and unknown evils would befall the boat and its occupants in his
absence.

He often spoke of the capricious *djinns* who fly with the sea
winds, to bring mischief and to create chaos amongst the incau-
tious, the unfaithful and the unaware. He boasted often of episodes when, in his opinion, only his presence had exorcised
dangers. Once the boat had not capsized when he and Paolo
caught three marlins in quick succession; only they had suc-
ceeded in getting enough bait that day, and in finding the yellow-
fin tuna in August, when the fishermen who are worth their salt
never venture too far to tempt fate beyond the coral reef –
because the monsoon blows with such rage that all fish seem to
disappear, and only the giant squids of the depths remain, and
the lobsters which have no soul.

Unworried by these predictions of gloom, Paolo and his
friends went out happily, leaving Ben still standing in the doorway, shaking his head and mumbling. I went to Mombasa to buy
kangas and baskets at the bazaars.

When I returned I could feel something had happened even
before I got out of my car; no one came to greet me. There was
nobody around. From the empty kitchen door a cat looked at me
impassively. It was a strange cat, which I had not seen before.
The friend who owned the Kilifi Plantations, and was a true mine
of all coastal legends, had told me the Giriama never chase away
a stray cat begging at their door, as they believe it is the returning
soul of a departed person. With this thought in mind I went to
look over the cliff.

Down on the beach, I could see a small crowd gathered on the
shore. Surrounded by the children, our friends, the staff and quite
a few passers-by, Paolo was crouching close to the biggest fish I
had ever seen. Its white upturned belly offered to the sun was
interrupted by the ugly line of a mouth like a trap, set with curved

teeth. A triangular fin protruded from its grey back. Defenceless and inert, it still appeared dangerous.

It was dead and it was a shark.

And so came the story.

From the depth of Vuma, after a full morning of fishing, Paolo had swum back to the surface, his prey, a *cole-cole*, still alive and moving, tied at his belt. Before emerging, the instinctive alarm which saves the life of the hunter warned Paolo to look down; and below him, from the ink-black darkness, a fish was approaching fast, looming bigger and bigger, as if magnified rapidly by an invisible lens.

When it was a few yards from Paolo it stopped and shivered, collecting itself before springing to the attack with a jerk. The snout slid back like a mask, exposing horrid long fangs in two circular rows; the small round eyes focused on Paolo, cold and expressionless.

It was a bull shark, of the kind which attacks more often than any other. The one known to eat sailors.

Too late to climb out of the water, too late to let his wounded fish go. In the blue, alone with the shark, Paolo had lost no time to think. He grabbed his loaded spear gun, placed it between his legs and pointed it straight at the shark's head. At the moment it was about to strike, he let the shot go.

The head of the spear penetrated the shark's forehead at a right angle. The shark stopped, writhing, and with overwhelming relief Paolo watched him sink. He sank like a stone, growing smaller and smaller, almost now out of sight in the depths, pulling with him metres and metres of rope and the floating balloon attached to it.

Elated, adrenaline pumping high, safe, Paolo climbed into the boat, from which his friends had been watching in breathless alarm. Together, with great difficulty, they hauled in the huge rigid beast. He was almost as long as the boat, heavy, his snout still contracted, his glassy eyes expressionless. They forced his mouth open, exposing the fangs, and fingered them, exclaiming at their size. Roping the shark to the boat so they could tow him back to show us, they prepared to start the engine.

As for buffaloes and for lions, there is always another life for a shark. Before the engine sputtered and ignited, a long shiver shook the fish. The wrinkled snout relaxed, and he struggled to loosen his bonds. Then he started to swim off with immense strength, pulling the dinghy, and Paolo, and his friends with him. A scene out of *Jaws*.

Here Paolo paused, looking around for effect.

On the Kilifi beach at high noon, with the reef shimmering and sparkling on the horizon, the tiny waves of low tide gently lapping the shore, and the palm leaves rustling in the breeze, we could have heard a coconut fall on the sand. Paolo's audience waited without breathing.

He continued. In the fragile rubber dinghy, pulled wildly over the sea, pandemonium ensued, as we could all imagine. Finally, with some difficulty, they managed to recover themselves; the engine sprang to life, gained momentum and pulled the boat in the opposite direction.

Weakened, gradually the fish gave up. Dragged backwards, the water entered its gills and it drowned. It took them hours to drive back on a sunny, tumultuous sea, with their heavy cargo not quite dead and their boat made of rubber.

Nobody could tell a story better than Paolo. In his words the adventure took on the colours and sounds of an epic sea saga; Homer and the sirens of myth paled beside this real-life drama. The crowd and I hung from his lips, spell-bound.

Finally the shark was heaved laboriously on to the back of a pick-up and driven in glory to the Mnarani Club. The fish scale there, good for marlin and sailfish, proved to be too small, so with a cortege of supporters Paolo and his friends went on to the farm scale at the Kilifi Plantations.

The shark weighed 532 pounds. Someone took several photographs, one of which appeared in the *East African Standard*, with the caption: 'Mr Paolo Gullman, a fisherman from upcountry, with a 532 lbs man-eater Bull shark.'

'Quite remarkable,' commented one of the orthodox fishing club members, mumbling into his pipe. 'What a shame he did not use a rod. It could have been an all-Africa record.'

This adventure became the talk of Kilifi for weeks.

The only person who did not seem impressed was Ben. '*Kama mimi alikua huko nikushika samaki, hio papa awessi kukaribia,*' he grumbled, tossing his cap. 'If I had been there, that shark would have never dared to attack.'

Nobody cared to deny it, and perhaps he was right.

The Story of Nungu Nungu

For Gilfrid

They left a great many odd little foot-marks all over the
bed, especially little Benjamin.

BEATRIX POTTER, *The Tale of Benjamin Bunny*

In the early days at Laikipia, I decided to carve a vegetable
garden and an orchard out of a shrubby area behind the
house. It was close enough to the kitchen to ensure some sort of
protection from the various pests which would attempt to eat the
produce.

Elephant loved bananas and oranges, gazelles lettuces, spinach
and broccoli, moles fennel, potatoes, carrots and all the tubers.
An astonishing variety of little vermin devoured just about anything, and birds wiped out everything else, including, thank God,
insects.

An ingenious and noisy – if primitive and rather messy – con-
traption of tins rattling on poles, vibrating strings, floating net
and long strips of plastic that flapped in the breeze was devised
to discourage the birds. It was set up, tested and discarded in
turns. It was supervised by a formidable *spaventapasseri*. Dressed
in one of Paolo's old jackets and topped with one of my mother's
forgotten straw hats, it conducted, like a veritable wizard, this ill-
assorted orchestra from the height of an old broomstick.

Nasturtiums and tagetes were planted around the vegetable
beds to discourage the flying insects with their effluvia; ashes
from the fireplace were scattered round tomatoes and courgettes,

and hay round strawberries, to deter crawling bugs; while a wire net, practically dik-dik-proof, was set on posts round the whole compound.

At night the Tharaka *askari* stood guard with a primitive catapult, of ancient design and time-proven accuracy – identical, I suspected, to the one David had used to kill Goliath. Like his biblical predecessor confronting the mythical giant, Sabino, creeping in the shadows, knocked round stones smartly on the backs of any elephant approaching the guava. Outraged trumpeting piercing the night meant he had hit the target, and became a picturesque feature of our evening meals, and caused a bewildering entertainment for our European guests.

The elephant did not seem to mind.

One night an extraordinary commotion made us run to look. A young elephant, one of a group of fifty which had lately been foraging on my bananas, had fallen into the septic tank. His companions were pushing him out. As always on these occasions, nobody could find a camera, but next day we ordered an electric fence.

A wire mesh cage was eventually erected to keep away even the most daring mouse birds, and my vegetable garden became an unconquerable fortress. Yet . . .

'*Muivi alikuja kukula mboga.*' 'A thief has come to eat the cabbages,' the gardener Seronera announced mournfully one morning, holding out to me a half-eaten cabbage leaf. Cabbages took ages to grow.

'He dug a tunnel below the enclosure,' he offered by way of explanation. We had not thought of this. He shook his head in knowing admiration.

'What! That thing is quite clever.' He produced a long quill, striped cream and brown, grinning.

'*Ni nungu nungu: yeye napenda mboga saidi.*' 'It is a porcupine: he is wild about cabbages.'

I went to look, and sure enough, in the crumbly fat earth,

lovingly manured and watered, little prints like a toddler's hand marked the soil. A large hole had been dug below the netting, and the tracks went inexorably to and from the cabbage patch. The thief had helped himself liberally. Another quill, like a signature, was stuck in the softened earth. It was a porcupine sure enough.

The net was repaired, but a few mornings later, Nungu Nungu came again. Another few cabbages were munched away. The little child's marks told the tale. This time we dug the netting deeper into the soil. It did not seem to help. Some weeks, and quite a few cabbages later, I decided to catch the thief with a trap. We already had one.

Nguare and Lwokwolognei had built it, to catch a leopard that had been killing sheep and which we let go in Samburu Park. The leopard had not liked being trapped, and had snarled furiously from behind the canvas with which we had covered the cage to protect him from the light and the unsettling sight of humans. How deep had been his bronchial roar. How much its sound had been for me the voice of Africa, and of all the unknown, untamed world around me.

This cage was sturdy, made out of thick timber, with heavy duty wire sides and a trapdoor, craftily connected to a bait. The door would slide down with a bang, capturing the thief when he took the lure. Formidable, I thought, for a porcupine. In the leopard's case, the bait had been the almost rotten carcass of a sheep. This time, it was a cabbage.

We set the trap meticulously, careful not to leave any human scent to reveal the plot to Nungu Nungu's suspicious mind. The most inviting fresh cabbage was placed in the middle. Nungu Nungu did not resist. Next morning, he was there.

Nungu Nungu was huge, with brown liquid eyes. He was covered in long quills which stood out, erect, while, at our approach, the hollow ones of his tail made a curious dry noise of frantically shaken castanets. He was pacing his prison, trailing his spiky train with the hauteur of an outraged great Red Indian chief. The closer we came, the louder, more threatening, the noise seemed to grow. But, all in all, he appeared remarkably quiet for what must have been a disconcerting experience.

And he had eaten the cabbage.

We put the cage on my pick-up, not without difficulty, and drove off. I stopped in a bushy area several miles away. With the help of all the gardeners, our tracker Luka and a grinning Emanuele we lowered the cage with great caution in the shade of some shrubs. Then I opened the door, and walked some distance away.

After a time, feeling safe, a little face peered out, and we watched Nungu Nungu zig-zag rapidly off through the undergrowth.

A few nights after this episode, the cabbages were eaten once more. We set the trap, and again, we found a porcupine inside next morning.

How could we tell porcupines apart? Seronera suggested it might be Nungu Nungu's mate; perhaps there was a family of them?

But Colin, my ranch manager, said it would be the same one, who had come back. Leopard, like some domestic cats, were known to have walked hundreds of kilometres back to their territory. It did not seem possible, yet who could know what instinct might not guide this creature back home?

This time I sent the pick-up to release him further away still, well over the Ol Morani boundary, on a vast grassy *mbogani* dotted with low *carissa* bushes, a long, long way even for a determined Nungu Nungu.

'*Kwisha rudi.*' 'He has come back,' announced Seronera some while after.

Was he really the same one? It was hard to believe.

Driven by curiosity, determined to solve the riddle, I found a tin of paint, and, through the wire mesh, sprayed the crackling quills a vivid green.

This time I drove the cage up to the Pokot boundary, with the usual assortment of giggling spectators in the back.

At sunset, when the shadows were long, and the flocks of guinea fowl settle on the highest branches of the acacia to sleep, and the swifts dart low, piercing the sky with their screeches, and the tree-frogs wake up with a sound of fresh bells, we set a green Nungu Nungu free over the boundary line.

'*Kwisha rudi tena*,' murmured in awe Seronera a week later. He held a long quill, partly covered in bright green paint. I laughed and laughed. We all did. Such persistent, pig-headed greed commanded respect.

After all, I never liked cabbages. We tried artichokes, instead.

Elephant Ballad

For HRH Prince Bernhard of the Netherlands

Recognizing that a creature of another species is in danger
from one's own kind; going to the aid of that creature . . .
imply the exercise of true compassion and also other most
sensitive emotions.

IVAN T. SANDERSON: *The Dynasty of Abu: A History and*
Natural History of Elephant and Their Relatives,
Past and Present

Its breath is said to be a cure for headaches in man.
Cassiodorus, *Variae*, X, 30

T he man limped towards my car holding on to rudimentary
crutches made of cut branches. Below his loose turban,
feverish eyes peered at me above gaunt cheeks.

'*Jambo!*' he addressed me shakily. '*Mimi ni ile ulikufa mwaka*
hi. Unakumbuka mimi? Ulitembelea musijana yako na ngamia.' 'I
am the one who has died this year. Do you still know me? I used
to bring your young girl riding with my camels.'

Of course I remembered him. His name was Borau, a camel
handler of the Boran tribe, whom we had employed in Laikipia
for years. He herded camels day after day, and often came up to
Kuti to hold the bridle of Sveva's camel when, at four or five, she
had developed a passion for camel riding. He spoke constantly to
his camels in the ancient camel language developed over generations and generations of close relationship with these extraordinary creatures, the noblest of African livestock and essential to
the survival of his tribe and the related Somalis.

'*Toh-toh galla.*' The camel sat, crashing down on its knees.

'*Oh. Ohohoh oh galla.*' And the camel went.

'*Ahiaeh ellahereh.*' And the camel drank.

'*Kir-kir-kir.*' The camel trotted faster and faster . . . and so on.

Day after day, off went Borau at sunrise to guide his herd to the grazing grounds. Until one day he met the elephant.

A morning like all September mornings; a sky of deep rose and the stillness of dawn; the profile of the horizon black, with sculpted acacia trees; birds chattering from the *lelechwa* shrubs; and a yellow sun, round and flaming, rising in a glow of promised heat.

The camels had waited patiently, chewing their cud with a crusty noise of long worn teeth, sitting on knobbly knees and surveying through sad eyelashes the morning preparations in the *boma*. A hot mug of spiced tea to wash away sleep; a bowl of sour camel milk; incitations, calls, and they were on their way. His camel stick firm across his shoulders, today Borau headed towards Marati Mbili.

He liked his job. He knew nothing else but walking in front of the camels, timing his agile step to their rhythm, curiously similar to them in his long ambling steps and his thin legs with heavy joints, built to march without pause. Or walking behind the camels, following their large soft feet that raised no dust, and left only neat rounded prints, like the shadows of a leaf.

He knew their favourite browse, and interpreted their needs like all herdsmen who match their lives to those of the animals they tend. His very existence was camels.

Today was drinking day for the camels, and Borau decided to wash in the *marati*. The camels drank first, extending their necks to the troughs, stimulated by the Song of Water, a triumphant biblical lament as ancient as the need to drink.

'*Hayee helleree, oho helleheree.*'

Immediately after drinking, the camels started grazing, nibbling with prehensile lips on nearby bushes and filling their discerning mouths with *carissa* leaves. The sun was higher now, and Borau slid off his *shuka* and his head-scarf to wash.

It was then that a big young male camel, who had been mauled by a lion, started a courting skirmish with one of the females, but was intercepted instantly by the dominant rutting male. The old camel came behind the suitor with frightening gurgles, and chased him off furiously. The younger one crashed away through the bushes with alarming speed, and was instantly lost to sight.

It is extraordinary how suddenly and completely the African bush can swallow animals. A shiver runs through the leaves, as the shrubs recompose themselves like ripples settling after a plunging stone. A cloud of dust suspended in the air; a whiff of rank smell; a sudden intake of our breath; perhaps the impression of a shadow, darting too quickly to let us focus on what we think we have seen. Only the prints of feet running on the track remain to prove that a herd of animals has just passed.

Borau sped after his camel, dressing while he ran. He tracked its rounded foot marks, but soon lost them in a mess of fresh elephant prints round a muddy water-hole. He looked and looked in vain; not only did the abundant elephant spoor confuse any other marks, but it announced the presence of a large invisible herd. It was wiser to go back to the camels, and make sure they did not become frightened and scatter in all directions.

Now, from the signs he saw, he knew the elephant were ahead of him. Not that it mattered. Borau was used to this. He just ought to be on guard, make sure he remained downwind, so that his scent would not alarm them, and move on light feet, hardly touching the soil, like the impala.

Soon he sighted the backsides of two elephant emerging from the sage just a few steps in front of him. He moved behind them carefully, all his senses alert so as not to disturb them.

He never heard the cow elephant which followed silently. He never saw her, until it was too late.

An instinctive glance over his shoulder. A large hovering shadow obscuring the sun for an instant, the pungent smell of ripe dung and hay, a hot breath fanning his neck and shoulders.

The look of a yellow eye fixed on him, from a few feet above. Large grey ears flattened against grey temples. Extended trunk curled up to expose long tusks. The horrible recognition that the elephant was after him, and that he could not escape.

Blind terror squeezed his heart and Borau ran.

In total silence, the elephant ran after him. She was a heavily pregnant female, young enough to be quick, agile, and gain deadly speed; old enough to remember that man is the only danger to elephant. Old enough also to have been part of a group caught in a poaching ambush, when the screams of pain and the smell of the blood of her fallen companions left an indelible mark in her memory. And female elephants are known to become over-protective, often touchy and aggressive in the time immediately before and after they give birth.

Borau ran and ran, mindless of thorns and sticks tearing his clothes apart, blinded by the sweat filling his eyes, and while he ran, he knew that he was going to die.

The thought of a slender girl, her velvet eyes laughing below her head-shawl; a bowl of camel milk steaming in the chill of dawn; the call of a child running towards him; the familiar hollow sound of the wooden camel bell. The simple things of his lost life now beyond his reach.

The earth vibrated, shaken by the elephant's feet, and by the thumping noise of his heart.

He wildly looked around for somewhere to hide, for a tree to climb. But there are no trees, in thick *lelechwa* country. Then the impenetrable, impassable *lelechwa* gave way to an open *mbogani*, littered with roots and boughs. One caught his foot, and he tripped, face down, onto the hardened soil, his nose squashed into the dust. With a jerk he turned and looked up.

The elephant was on him.

In perfect silence she went down on her kneees at his side, and in one movement lifted her tusks high and plunged them down into his leg. The tusks were butter-coloured but hard as spears, and like spears they penetrated his thigh like butter. The snap of fractured bone sounded like the snap of a broken branch. No pain. A spreading numbness.

The elephant stood, towering over him, looking down at his squirming body, as if to make sure he could hurt her no more. Slowly, deliberately, she lifted her foot above him. He screamed.

Startled by the strange noise, she stiffened, her foot hesitated and, in this pause, Borau instantly saw his chance and started pleading. If his camels understood, why not the elephant?

'Hapana. Hapana, ndovu. Wacha. Kwenda. Akuue mimi, tafadhali akuue rafiki yako.' 'No. No, elephant. Leave me. Go. Do not kill me, please do not kill your friend.'

Had the elephant cow ever heard a human voice before? The new sound pierced her open ears, puzzling them with a new note. She seemed to listen. Her large ears flapped once, twice. Her foot came down onto the exposed face, but not to hurt. It stopped almost in mid-air, then descended to touch him.

Borau was too shocked to protect his face with his hands, and the elephant's nail caught his turban, and undid it. The material came loose, covering his eyes. The foot hovered over him slowly, deliberately, brushing down the length of his whimpering body, but pausing to feel his head and chest. He could see now the furrows dug into the sole of the elephant's foot by walking thousand and thousand miles over thorns and rocks. She probed him with surprising gentleness, as if the sound of pain and fear in his moaning voice was one she could understand.

After a while she stood back, and, more confident now, Borau agitated his hands and began calling out loud with all his remaining strength, the camel's command to run: *'Kir-kir, kir-kir.'* 'Go fast. Go fast.' He screamed louder and louder.

The elephant shook her head from side to side a few times, as if to chase that sound away stamping round him in the dust. Then she turned, and crashed away trumpeting. Only the cicadas remained, to fill the sudden silence with their eager songs.

The pain began to pulsate. Borau's mouth was parched and dry, his leg wet with blood and urine. He tried to move, to crawl towards the direction of the track, but he could not.

Perhaps the people back at the camp would notice that he was missing; when night came they would come and look for him and find him. But when night came so would the hyenas, and the lion, and the little silver-back jackals with their greedy mouths. If a sheep or a steer were lost, he knew it would never survive a night outside the *boma*.

The smell of blood, the smell of fear would attract all the scavengers. It was strange that there were no vultures yet; only the formidable presence of the elephant could have discouraged them, but he knew they would not be long in coming. In Africa there was always a vulture circling high, close to the sun, looking down with its telescopic eye for a dying animal on the plains.

The vultures would come from the sky, free-falling fast like bombs, and land on a branch, gathering their wings about them – first one, then another and another, until all the trees would be black with them. While the air filled with the sinister sounds of their presence, the vultures would sit and wait with undertakers' patience, and they would not have to wait long. Then one would come close, with awkward leaps, flapping his wings with raucous chortles of anticipation – the grotesque vulture that goes for the eyes first.

Visions of death and carnage filled Borau's mind – a sense of his own total vulnerability, crushed, unable to move, dying alone in the bush, an easy prey to any animal of the African night. He wondered at his own destiny, he who so many times had escaped malaria and the grip of high fevers, and infections and wild animals. Was it the wish of Allah that his road had reached its end in such a way?

The sun was setting now, he knew by the changed sounds of the bush; the sun he saluted every morning and to which he prayed every night. He tried to talk to Allah. Was God too remote from this *lelechwa* country?

He prayed for company, for any company. And he soon realized that God had listened.

He was not alone.

Slowly, through the fog of his total misery, Borau became aware of presences round him. They were gathering quietly, betrayed only by a noise of a broken branch, by a stomach gurgle, a shuffle, a deep breath, a rustle of leaves. They were extraordinarily quiet, and they were coming towards him. Their large feet did not hammer the ground. They waded through the bush with great ease and the calm of creatures who are unafraid. Soon their vast grey shapes cast vast shadows over him.

And Borau knew, without fear, that the herd of elephant had come back.

Earlier they had stopped feeding to watch what was happening. Now they came curious and unfrightened, to see what it was, this small trembling animal on the ground. First, came the young ones, tended by the matriarchs; they ran to him with open ears and stopped a few feet away, to observe him with attentive eyes. Then, one by one, the entire herd approached until they all stood round him, watching.

With feverish eyes, Borau looked up at the elephants as they looked down at him. He gazed into their yellow eyes, which examined him with benevolent attention, and he could feel that they would not harm him. In a weird way, he knew that, on the contrary, they would shelter him from the nocturnal dangers, and that for as long as they were there to guard him, no predator would dare to approach.

For a very long time, they stood in silence, as if studying him, and during that time, Borau talked to them. Heads lowered towards him, ears wide open, while they appeared to listen to the universal language of pain and surrender.

A trunk lifted, stretched, reached out and then another. Tentatively, smelling him and feeling him as gently as the caring hand of a nursing friend, they all touched him with their trunks. Quietly, they inspected him, carefully, unhurriedly, as if to reassure him.

Night had now fallen, with calls of guinea-fowl, grass-crickets, tree-frogs and nightjars. The elephant began to feed around him, like silent guardians. Now and again they each came back to stroke him. They ate, and then they came and checked, as if to reassure themselves that he was still there and fine, as if to reassure him that they were there to protect him.

Time drew on; Borau curled up on the cool spiky grass, trembling now with fever and shock, almost unconscious, but feeling utterly safe in their mighty protection.

They waited there around him while night approached. And who knows for how long they might have stayed. Even when the sound of an engine broke the silence, they still waited, alert now, with heads held high to smell the wind, ready to flee from the only animal of which they were scared. Car lights pierced the night and human voices; the engine droned closer now, advancing wheels opening up the shrubs.

Only then, like a school of dolphins going back to their ocean, leaving the shipwrecked sailor they have brought ashore to the care of a rescuing boat, did the elephant disappear, noiselessly, into the dark.

The Rhino That Ran
Fast Enough

And the experience has left me in some doubt whether a
rhino has such poor sight as it is commonly believed.

VIVIENNE DE WATTEVILLE, *Speak to the Earth*

I t is often difficult, in Africa, to surprise animals in the open,
as the lush vegetation or thick shrubs provide a myriad of
hiding places, into which the wild creatures, alarmed by the noise
of an approaching car, the crack of a twig under clumsy feet or a
whiff of your smell carried by a change of wind, can quickly dive
and disappear.

Often have I caught, out of the corner of my eye, a furtive
movement at a bend in the track, the shape of a tail or flickering
ears in the tall grass, and a shadow has dashed away faster than
my mind could register it, leaving an impression, only, of the
passage of some elusive life that challenged my imagination. Yet,
if I stopped to search, scanning the sand or the dried mud of the
game trail, I would find the unmistakable imprint of a large hoof
or of a paw, like a signature unwillingly left by fugitive feet.

The element of chance involved in crossing a path at the very
same moment as a rare or shy animal, and of seeing it for an
instant still in the sunshine or trapped in the car's headlights,
before the savannah grass or the night swallows it, has never
failed to puzzle me. A few seconds earlier, or later, and the scene
would have been missed forever.

I can recall countless such episodes, but perhaps the most

extraordinary of all was an amazing encounter, witnessed with the same sense of wonder by four pairs of eyes. It resulted, through its perfect timing, in the fortuitous rescue of a very rare little creature in distress.

One July morning in Laikipia, when Sveva was about five years old, I drove her down to Centre to fetch her young friend Andrew, Colin and Rocky Francombe's son, to come and play in our house at Kuti.

The air was still and hot, and the sky hung close, the colour of lead, as it is during the season of the rains, when the Highland winds stop blowing for a time, and the only movements are the tremulous flights of white butterflies, migrating westward in endless clouds of palpitating wings. They go, like waves, incessantly in the same direction, as to a rendezvous they cannot miss at the far end of the horizon. The golden air seems full of the snowflakes of an improbable summer storm, or of petals from a creamy bougainvillaea scattered in gusts by an invisible breeze.

The transformation the rains brought to Laikipia was always breathtaking. My car skidded in the fresh red mud. The tufts of new grass at the sides of the track were emerald green, and clusters of frothy flowers covered the *carissa* shrubs, mixing their intoxicating jasmine scent with the sweet and warm perfume of the flowering acacias. The wild animals seemed to be reborn, well fed, glowing with health – frisky impala, shiny waterbuck, fat zebra and placid elephant foraging unhurriedly from the taller branches. The red dust which had been dulling the bush like clinging rust had vanished, leaving shining grass leaves and fresh buds. With the children chatting in the back, I was driving slowly back to Kuti, concentrating on the beauty and bounty of the wet, sumptuous African landscape.

It happened so suddenly; I was so unprepared. A small grey something darted across the road, almost running straight into the car; a strange little creature like a cut-out cartoon, invading by some trick of imagination the real world around us.

It was a baby rhino, no taller than a dog, dashing in front of me with amazing speed. Alone. Its eyes were fixed on the track ahead, but on perceiving my car they turned towards it and, for a fraction of a second, I could read in them, with great surprise, a look of pure terror. For whatever reason, the baby rhino was scared to death, and I realized that he was running for his life.

He passed inches from my bumper, and in a moment he was gone. I slammed on my brakes, and the car came to a halt, skidding. In the very same instant, another car, coming from the opposite direction, stopped in front of me. In Karanja the driver's face, I read the same astonishment. We looked at each other, and I jumped out of my car to see.

There, standing in the middle of the track, a few metres behind my car, was the smallest black rhino I had ever seen. His skin looked soft and smooth, like a rubber toy. On his nose only an insignificant protuberance indicated where his horn would one day grow. His eyes were tiny, porky and concentrated on me or, rather, on my car. Absurdly, but surely, I caught in them the reflection of my own surprise, which had wiped off what had been his expression just seconds before: unmistakable, overwhelming terror. And then a definite look of relief, oddly of recognition, of joy almost, flooded into his piggy eyes, as if the encounter had in some weird way comforted him.

An instant later, quite unexpectedly, he began running towards us, aiming straight for my open car door. I did not move, but suddenly my scent was brought to his sensitive nose by a change in the breeze, and human scent meant danger. Startled, looking betrayed, he came to a sudden halt. His head went down, a snort came through tiny nostrils, a comic determination born of instinct took over. He charged and, before I knew it, he knocked his embryonic horn against my bumper. It was so funny that I burst out laughing, and so did the children, whose bewildered faces, with mouths opened in amazement and round eyes which had missed nothing, were pressed to the back window. The noise startled the rhino; and in one movement he turned on his heels, swerved suddenly and trotted off faster than ever through the undergrowth.

Only the bush was left and the empty road, where a faint cloud of dust soon settled.

I turned to Karanja, my other witness, to comment on the event: his mouth, too, was agape, and his eyes widened incredulously. There was more to it than had met my eye. He had been driving a tall land-cruiser and, as the scope of his vision was much wider than mine, he had seen what I could not. His fat hand came out gesticulating excitedly and indicated a point on the side of the road I could not see. I noticed that he had difficulty in finding his voice.

'*Simba*!!' he screamed at me finally. 'Lion!'

'*Iko simba uko nafuata hio mutoto ya faru.*' 'There is a lion there following that rhino child.'

That explained the desperate fear. I turned, stood on my tiptoes, and sure enough, through the tall grass and low *carissa* shrubs, I distinguished the yellow shape of a stalking animal. A moment, and it was gone, leaving the tracks of its claws embedded in the hard soil.

The rhino had disappeared too. His hour had not come yet. Our presence at that precise point in the road at that very moment, had, by some arcane design of fate, saved his life, just in the nick of time. I wondered for how long the chase had been going on. Where was his mother?

Karanja had the answer: 'It is the rhino child whose mother was killed by poachers.'

Why had he run towards me? I thought about it for weeks, and asked all the animal experts I came across for an explanation. Surely he did not come to me to protect him. He was a wild rhino who was not accustomed to humans. But I had been driving a low off-white Subaru, splattered with mud.

Its size, its colour and its shape were familiar to him.

My car was the closest thing to his mother the rhino child had ever seen.

Night of the Lions

A brave man is always frightened three times by a lion:
when he first sees his track, when he first hears him roar
and when he first confronts him.

> Somali Proverb, from *The Short Happy Life of Francis
> Macomber* by ERNEST HEMINGWAY

It was the *askari* who brought Meave back in a wheelbarrow, one April morning when I had been gardening.

The bougainvillaea were a riot of purple and needed taming. The first long rains were just due to begin. It was the right time.

I loved the concentration, which left my mind free to wander, thoughts forming, stories unfolding, solutions finding shape. I liked the sun on my face and the sudden breeze and the call of turaco from the treetops.

From the shade of the large yellow fever tree I saw the strange procession approaching. Through the pepper trees, in and out of sun blades, they came towards me. I stood, putting down the secateurs I had been holding, and touched the warm head of one of the dogs. There was a weird gloom in the way they moved.

Dragging his long *kaputi** on the ground, his cap askew, preceded by two of the other dogs and followed by a gardener, Nyaga pushed a handcart from which the matted tail of Meave, caked with blood, trailed on the grass.

The dogs approached cautiously, sniffing with lifted noses, baffled. Meave was the oldest female of my Alsatians, the leader of the pack.

*Military coat worn by night watchmen.

Nyaga looked up at me apologetically, his normally happy face split in a thousand wrinkles:

'*Pole. Ulipata yeye kandu ya bara bara. Aliona ndamu kwa nyayo: ni Simba. Uliona mugu.*' 'I am sorry. I found her on the side of the road. I had seen tracks of blood on the path. I saw the footprints. It's a lion.'

Nothing I could do would save Meave.

She was alive; motionless; she looked up at me with a stoic acceptance of her fate, black eyes still, ears alert, impassive and fearless as Meave could be. I looked down at her twisted body, her hindquarters seemingly dead.

I asked my cook, Simon, to bring a basin of warm water and ran to fetch my first-aid box.

I washed her gently and shaved her rump. She did not whimper, as if she had become insensitive to pain. And excruciating pain there must have been because, on the bluish skin my razor exposed, were four deep gashes from which blood still poured, round deep holes spaced like the fangs of a very large lion.

Flashes of the scene darted through my imagination. There were so many possibilities. Perhaps the dogs, out on their night's escapade, had found the lion on the kill. My dogs have learned that the lion's presence means meat to scavenge. They follow the feral scent through the bush early in the morning, before sunrise. They come back panting, wildness in their eyes and a rank smell on their fur, and jump into the fish pond exhausted, before drinking noisily from their bowls.

The story of their night's encounters, of meals shared with jackals, of bones fought over with hyena, the excitement of the hunt, will remain a mystery. I could imagine a short fight, the lion's annoyance, a deep growl . . . a snarl that becomes a bite, deep, deadly, effortless. A crunching sound of fractured bones. Like a rag, Meave flew in the air, landed in silence. The dogs scattered. The lion went on eating, unaffected. Meave's back was broken. She managed to drag herself a few metres away, below a *lelechwa* bush; and there the *askari* found her.

I disinfected her wounds with red mercurochrome which mixed with her blood, blotted them dry and stitched her thick skin with the curved needle from my kit. I applied powder and bandages, I did what I could: but I saw that her spine was broken, her back legs paralysed, and knew that the lion had won. I put her on a mat, in a quiet corner near my room where I could keep an eye on her; she drank some milk, but I could see that she was incontinent, and after two days she refused food and lay there with closed eyes, barely breathing.

It was a gun shot at sunrise, at the bottom of my garden.

I had asked Aidan to help do what had to be done, but which I was incapable of facing.

She had been sedated with Nembutal and had fallen asleep, while I caressed her head. The echo resonated in the paling sky, startling the birds. A flight of starlings took off from the fever trees with a vibration of wings in the sudden silence.

It was a fresh dog grave next to many others, not far from the place where Paolo and Emanuele rest.

I went to look at the tracks that same morning and saw the lion's pug-marks in the dust. They were soft and rounded, immensely powerful and emanated danger.

Even more than a direct encounter, even before you hear his roar for the first time, it is the lion's footprint, marking the dust of a game trail, the unmistakable proof of his passing along the very path you are walking, that makes you feel that you are really in Africa.

After a time, living out in the African bush, you learn that there are animals you see and others whose presence you sense, not far from you, hidden by the undergrowth at the side of your track, the thick *lelechwa* or the clusters of euclea. Or in the night's shadows, expanding impenetrable out of your windows. In the border of darkness left outside your tent by the embers' glow of last night's campfire.

The elusive ones, whose scent only you detect, brought by a change of wind, betrayed by the cracking of a twig under a cautious paw. A powerful breathing, rasping and alien, rhythmic and implacable, like the very essence of Africa.

Those, the animals you never see, are the ones that give you the deepest shiver of emotion, the sharper taste of fear, the strongest longing. They are the ones you will remember.

Like the nights of the lions.

Not so long ago, in January 1997, when Sveva was about to return to school in London, she asked me if she could spend at The Nest one of her last nights in Laikipia, before returning to Europe and the cold winter ahead. She hoped to see rhino. She would go on a night picnic organized by Shahar, our young agriculturalist from Israel, at the hot springs below the Retreat at Maji ya Nyoka, and would be dropped at The Nest, where I would fetch her next morning. She brought her sleeping-bag and her hurricane lamp and of course a torch. She would have a radio handset. She had done it before. But I was uncharacteristically uneasy, and went to bed, myself, with a handset radio resting on my pillow.

Even our cook, Simon, seemed doubtful tonight and looked at her seriously.

'*Makena,*' he pleaded, '*hapana sahau radio. Chunga sana.*' 'You, The One Who Smiles, do not forget your radio. Be very careful.'

Lions had been roaring at night in Laikipia in recent days, more lions than I had heard in a very long time. At night I was often woken by my dogs' growls, by their rush of barking breathlessness as they ran for the door of my room where they slept, and were quickly lost in the night. They scattered on the lawn, howling high, as they felt the lion's presence on the hills. But when lions came close, the dogs kept quiet. They cowered back to my room and lay down on my carpet, with alert ears, unsleeping.

This night was no exception. The rhythmic deep growls echoed on the hills on the side of the old Boma ya Taikunya, near the Ol Morani boundary line, and came closer. I could almost hear their padded feet on the track below Paolo's dam, along my garden, outside the guest rooms, as they moved towards the water trough of the *menanda*, where they could find buffalo, eland or zebra. I let the water overflow especially over the rim of the tank, where only elephant could reach it, so that it collects below it in a muddy hole, a delight for the buffalo. I suspected that lions and other carnivores had become increasingly attracted to the area. Landing one afternoon, I had seen a cheetah standing at the end of my airstrip, on an old anthill, checking out some young impala, undisturbed by the noise of the plane and swirls of dust; and at night I could often hear the magnified, see-saw purring of leopard: many impala went to the water.

I found it hard to sleep. Memories of danger haunted my mind. Still fresh was the memory of the recent encounter with the calf killer, in the last night of full moon.

When the moon rose I could see the cattle sitting around the *boma*, not far from the embers of the *wachungai*'s fire. They were mostly white, the cows of the breeding herd with their small calves next to them.

I was watching them through the thorny branches of the shrub I was sheltering under. My head covered in a cotton shawl rested on a double pillow, my blanket under my chin against the chill eastern wind. I was alert, and waiting. Next to me, stretched fully dressed on the mat, were Aidan with his lion rifle, Issak, the headman with a spotlight, and Ali, the head of the *wachungai* at the *boma*.

The lion had come every night for the last two weeks, most nights taking off with one or two calves. The lion had developed a passion for newly born calves. With any amount of young buffalo, fat zebra and tasty eland to feed upon, a lion addicted to killing cattle becomes a vermin no rancher can afford. Every week I had the cattle's *boma* moved out of the lion's territory. But the lion had followed, lured by the ease of taking tender young calves. Thirty calves had been taken, and I was left with no alternative.

Tonight, if the lion came, he would be shot.

He would come in the dark of the moon, at the time when the cold bites hardest, when the embers are greying and the men asleep. Wrapped in their *shukas,* curled around the campfire, they would sleep without dreams until dawn, apart from the one keeping watch, fighting off sleep with scalding mugs of spiced tea, smoky with honey and milk.

The noise of cattle was in an odd way soothing, despite the snorting ruminant gurgles. The smell of fresh dung was overpowering, herby and natural in an odd way. I liked it.

I reflected that the smell had not changed in Africa over the centuries: no artificial fodder had been introduced to pollute the old ways. Cattle here lived outdoors all their lives; they wandered freely for their food and water; they were not stabled livestock, tame and safe in cosy enclosures where everything was available but freedom.

Here at the equator they got wet in the rain and walked for miles in the scorching sun to find water. They were resilient, patient animals, who knew how to survive; they recognized the song of the tick bird, the taste of new grass, they could smell a salt lick from great distances, and they knew lions.

It was as if a sudden wind had come from the *mukignei* shrubs in a short, intense burst; the same sound, like an immense, collective intake of breath. The cattle stood all at once, silent, alert, like weary ghosts.

And from the east, in stealthy, focused leaps, aiming straight for the youngest calf, came the lion.

Next to me, with astonishing speed, followed by Issak and Ali,

Aidan had sprung to his feet, and while I kicked off the blanket, thankful for having kept my shoes on, and rushed for my gun, I heard the shot.

It shattered the night and echoed in the valley, with the deadly, disruptive finality of a rifle's burst. Deafened for an instant, running after the others in total silence, I heard only the drumming of my heart. A few metres away, in the dust the lion rolled, doubled up, gurgling its hollow death rattle. In a few seconds he lay still, and his eyes, no longer seeing, were a phosphorescent pale green, like dying fireflies.

Driving home later, alone in the light of the moon, trying to avoid potholes and hares on the rocky road, with the dead lion in the back of my car, I had wondered at the strange ways of my destiny and thought of other stories of lions in the night.

Sveva was alone at The Nest. The Nest is my refuge beside Ol ari Nyiro springs, a shelter of wood and stone open to suns and to moons, under a roof of thatched papyrus, with no windows or doors. It is built on a slope, overlooking a salt lick, and it holds a thousand spells.

There is the spell of the wind and that of the gentle rain. The spell of the storm, of thunder and lightning in the livid equator sky; the swallows' and swifts' spell, and the starlings' which roost on the beam above the four-poster bed made from old twisted olives and stones from the bush. At night come the fluttering bats with their special bat-spell of ancient caves, and dormice peer curious from their tiny holes, running fearlessly to pick up crumbs from my bread. Genet cats and porcupine come, too, and leave their musky savannah spell, and in the night I try to guess which of them is chewing the remains of my guava. There is the spell that the orange-headed agama with the turquoise body laid on the stone under which Sveva's child hand had carved 'Paolo' and a heart of shells, and the powerful spell of the friendly cobra who sleeps underneath it.

There is the rhino spell and the elephant spell and the spell of the lone male eland. But the one that leaves me enthralled is Paolo's spell like a breath on my cheek, and Emanuele's, who had carved on to a log an acacia tree just like the one that one day would grow on his grave.

Presences of creatures and people like mysteries populated the shadows of my beloved Nest, but the strangest of all – and the one that kept me awake that night – was the one of the lion which sat on my bed.

A few weeks earlier I had visited The Nest with some guests, looking for buffalo. There had been a weird, sickening stench, the distinct feeling of a presence disturbed by our arrival, the glimpse, out of the corner of my eye, of a tawny shape dashing off down the slope, and in the sitting room, a large, evil, reeking excrement, which I recognized for a lion's. Never before in Laikipia had a lion entered a house.

Narumbe, my wildlife ranger who was our escort that evening, had confirmed it:

'Simba!' he murmured, shaking his head in awe. 'Simba ndany ya nyumba!' 'A lion in the house.'

He turned his proud head round, still, like a wooden sculpture, flared nostrils sensing the wind, his perforated lobes, pendulous at the side of his head, alert to detect a revealing noise.

The Nest is built high on the hill, a perfect spot to survey the valley and watch wildlife. A few days earlier a young eland had been killed by a lion, just down below. Now, a herd of buffalo was emerging from the lelechwa.

'Yeye natega mbogo hama siruai.' 'He would be stalking buffalo or eland.'

We had gone in, a sense of disquiet and unease. On the printed counterpane of my large four-poster bed, a fluff of coarse yellow hair confirmed the tale.

A lion on my bed. A lion on my bed, no one would ever believe it. That episode, and the knowledge that animals mark their territory with dung and urine, had taken some of the shine from spending nights alone at The Nest, a place all open, impossible to lock, like a verandah. Sveva had shrugged off the eventuality of another feline visit:

'It will never happen again. Really improbable. We have disinfected and cleaned everything. His scent has disappeared. He would hate the smell of antiseptic.'

She was quite right. I did not give it another thought. Not until now.

Aware that Sveva was alone at The Nest, I was disturbed, inexplicably alert and awake. I kept tossing on my pillow, resigned to a disturbed night of wakefulness.

The crackling of static that indicates that someone is fidgeting with the button of the radio microphone is as ominous as a phone ringing at night.

It started suddenly but hesitantly, and, although we have many radio sets on the ranch, I thought instantly of Sveva and was immediately awake.

'Who is at the radio?' I whispered in the microphone. 'Who is calling?'

It *was* Sveva. A faint murmur, tainted with fear and urgency.

'Mummy, there are lions. Lions. Close, very close. Help. Send a car quickly. Please help.'

I was out of the bed before she had finished talking, fumbling to get the lighter and light my candle. I grasped the torch, searched wildly for my glasses, dashed for the door, at the same time mumbling into the radio urgent encouragements and reassurances, suggestions of what to do.

'Take your hurricane lamp, ready to throw it if they come in, scare them off at the last moment, plan to startle them, do not panic, do not run. Whatever happens, do not run. It's going to be all right, I'm on my way.'

I was running. My heart beat in my mouth, I could not afford to lose time. I had run that corridor before, in a time of terror.

In my nightdress, barefoot, I ran again on the bare stones, wet
with night dew, over the wet cold grass, my dogs, puzzled, around
me. I could still hear lion roaring down in the valley. The *askari*
caught up with me as I fired the engine of the first car I reached,
the large land cruiser we use for going around with the guests. I
told him to stay near the radio in case I needed help.

'Simba!' I called out as the car moved off in a swirl of dust.
'*Makena, na itua mimi. Simba na-ingia Nyumba ya Mukutan.
Unaenda kusaidia.*' 'It is The One Who Smiles. A lion is getting
into the hut on the Mukutan. I am going to help.'

My voice sounded shrill, high-pitched, strangled. I was thinking about lions.

Lions are curious creatures, lazy, they fear practically nothing.
In Laikipia I saw them countless times, mostly at dusk or returning home to Kuti at night, in my headlights. They stand,
undaunted, in the middle of the tracks or beside them. They
move away slowly with an easy muscular roll of the powerful
shoulders, yellow and tawny, in and out of the undergrowth of
sage bush and acacia shrubs. They sit down abruptly, with sudden
abandon, attentive but not wary, yawning unhurried, never moving their inscrutable eyes from the car.

They often follow if I slow down, intrigued, unthreatening.
But their indolence is never to be mistaken for friendliness, for
wild lions are unpredictable to say the least. This is what I feared,
their curiosity, the potent, careless, deadly playfulness of cats.

Never before were eight kilometres covered in less time. I was
racing. Through the shrubs, into potholes, over stones and gul-
lies. Up the hills, down the slopes. Zebra scattered. Then the *lelechwa* thickened, the track turned sharply left where an old twisted
olive grew, the narrow trail climbed a rise, descended again: I was
there. In the headlights my familiar Nest stood ghostly and grey
like Dracula's castle in a Gothic tale. The wind blew sinister,
sending shivers through the bleached wild sage. The night behind
it held dangers and nameless threats. Where was my baby?

I hooted, frantically, with every blast of the horn scaring off my monsters. Sveva emerged tousled, frightened but intact, folding her sleeping-bag about her, holding out a tremulous hurricane lamp which looked so diminutive that, on second thought, no lion could possibly have been deterred by it.

She had, characteristically, recovered.

'So sorry, Mummy. It was horrible. He was so close I could have touched him. I didn't know what to do.' She have me a half Makena-smile.

'How did you make it so quickly?'

She shook her hair. In her Paolo-blue eyes I noticed a sudden glint.

'But don't make so much noise. You'll scare the rhino.' I knew she was fine.

Memories gather in the night more easily, associations come naturally. It was when I was crossing the junction to the research camp, with Sveva sleepy in the car, that my thoughts went to another night, well over twenty years ago, more or less on that very same spot.

A story of other lions, in other times.

I had left Rocky and Colin Francombe's house – the managers in those days – with Emanuele, my mother and Gordon, my dog. It was seven in the evening, dark already. It was the rainy season.

'Let's choose the middle track to Kuti,' I said to Emanuele. 'It is little used, mysterious, and there's more chance of seeing wildlife there.' I looked up at the darkening sky:

'It may rain later. The road shouldn't be too wet.'

In upcountry Africa one lives out of doors, and one notices the quality and the colours of the soil. There are endless varieties, and in Laikipia one could find samples of most of them. It is the yellow sandy dust of the savannah, or the marvellous grainy orange murram of the Niuykundu Dam; the fat, almost purple ochre of *mlima ndongo*; the brown forest earth of Enghelesha, fertile with humus; the grey volcanic gravel of the northern plains, and the light orange or white powder leading to Mawe ya Paulo, which penetrates everywhere, covering the leaves at the roadside with a sticky film. And, of course, the feared black cotton soil of *damu ya tope* and *maji ya faru*, which a shower of rain transforms into a messy glue which clings to shoes and feet and tyres until one gets hopelessly stuck.

There was a humidity in the air, a feeling of impending clouds, but the night was still clear enough after the day's downpour.

A buffalo crossed the track a few metres ahead, head high, wet muzzle towards us, and stood there watching us unafraid. I slowed down to avoid him.

'A big male. I must tell Paolo.'

His hind legs were encrusted with mud. The grey horns spread wide, heavy, worn in the middle with deep ridges, a sign of age. When the car was a few inches from him, almost touching his hide, he shook his head and turned on his hooves with the unexpected, lethal agility of old buffalo. In seconds he was gone through the *lelechwa*, splashing long spurts of brown mud.

The bottom of the ditch was flooded, and by now my car was too slow to negotiate the rise. Even in a four-wheel drive I felt the tyres struggling to grip the slippery bottom, skidding, burying deeper into the sticking clay at each burst of accelerator. It did not take me more than a few seconds to realize I could not get out until it dried out.

I went through the usual moves. Low ratio, reverse; I got out, wading in the mud and, helped by the others, put branches beneath the wheels; I tried again a few times but I was hopelessly stuck.

The light was fading fast. We had no radio. I quickly realized that Paolo, believing that we were having a drink with Colin and Rocky, would not think of looking for us for at least a couple of hours. Often Rocky would ask us to stay on for a bowl of soup and a glass of sherry and I would accept. Her cook, Atipa, boasted an endless repertoire of excellent soups, and I liked their company. Paolo would think we were having a good time. He might join us later. Only then would he realize that something had happened. In those days we had no internal radio network – and no way to communicate.

I looked up at the sky which was already dark, but with a luminescence at the horizon edge; I remembered that there would be an almost full moon tonight. The moon would rise in an hour, and even with the clouds there would be enough light to see my way back to Kuti. I would go with Gordon.

'I'm going to walk home; it's six kilometres from here to Kuti, I know my way; you stay where you are, keep each other company; I'll be back with Paolo in less than two hours. It will be fine. Cosy in the car.'

My words fell into silence. My mother looked at me, then at Emanuele. Nine years old, he was in Laikipia for half-term. He faced me, looking me straight in the eye, and even in the dark I sensed his determination.

'You are not going alone. I'm coming with you. Nonna can stay here.'

'If you go, I'm coming too.'

My mother sounded adamant. This was Africa after dark, a place full of wildlife. Perhaps she did not realize the danger or perhaps she did, only too well. My academic mother, used to museums and libraries, university lectures and hours writing at her desk in Venice. She did not want to miss out and had to be part of the adventure. If there was danger to face, she wanted to share it with us. I looked at her with new respect.

I argued for a few moments, but knew that I could not convince them to stay if they chose not to. Finally I surrendered. There was no time to lose.

'I will go ahead. You follow. Do not talk, do not make any noise. Stop if I stop. If anything comes too close don't run.' I looked at my mother's shoes and repeated: 'More than anything, do not run. Be quiet. Be still.'

I slip a pullover over my head. It would be cold. I nodded at Ema and moved on.

We were lighthearted to begin with, elated by the absurdity of the adventure. Gordon went ahead, sniffing, his tail erect like a banner. He turned to check on me now and again, came back to be patted on the nose, ran off again. Unlike the mad runs of our daytime walks, he kept close to us. We could still vaguely distinguish the silhouettes of the trees and the hills.

But I could sense my mother's justified concern and Emanuele's excitement. I felt the responsibility, and a mixture of comfort at not being alone and worry at having to plan for three of us, should we have an encounter with something large and unfriendly.

The inhabitants of the night were about.

It is amazing how the silence of the night in Africa becomes quickly animated with presences and noises, as everywhere creatures wake up and take over; from the valleys they come, from the thickets of euclea and rhus, from the crags, the savannah and the plains. Around and ahead and behind us we become aware of creatures moving. And yet, in their presence they are silent, betrayed only by a deep breathing, a stomach gurgle, the swish of leaves opening and closing as a large shape passes through. Sometimes it is the elephant: a snapped branch falling, splintered by a strong proboscis. A far-away hyena, crying to the moon from the hills; a whining jackal's chorus, a lion's roar, the see-saw of the leopard; but mostly it is the tree-frogs and the crickets and unknown insects singing incessantly in the background; and starlings when the moon is full.

So I walked, listening to the familiar sounds of the African night.

Sound of steps on cracking twigs. A disturbed guinea-fowl getting excited in a burst of protest.

We followed the narrow road, little more than a track. I could just make it out. In places the bush on either side was impenetrable and too close for comfort. Then the bush opened out and I noticed a change in Gordon. He stopped abruptly in front of me, a paw in the air, ears pricked, listening; his nose had caught a scent; a low growl formed in his throat, and I stopped in my tracks; Emanuele and my mother stopped with me. Gordon moved his head to one side, still growling, as if he could see and smell what we could not. Then the scent was caught by a breeze and Gordon relaxed and moved again ahead. We followed warily.

I heard sounds of buffalo ahead, a large herd grazing; the distinct noise of many mouths crunching on strawy grass filled the darkness, belching and shuffling, stones rolling under heavy hooves, a warmth and many bodies and green manure. They blocked the way ahead and stretched in all directions. They surrounded us. Gordon took charge. He struck a pointing pose, rigid, and attacked the darkness.

We all stopped, braced for a stampede, which happened, blissfully, in the opposite direction. The night became rocks rolling, shrubs crashing, loud mooing and running buffalo. Soon only their warm, acrid smell remained.

At Kati Kati tank there were elephant drinking. I had anticipated this, and the danger of unknowingly cutting across the path of a herd of elephant or buffalo going to drink. Gordon was silent and circumspect, cautiously moving ahead. I knew the wind was on our side. We managed to move on unnoticed.

There were about two and a half kilometres to go; I felt relieved; but Gordon's behaviour had changed again; he did not walk ahead any more; he stayed close to my legs, as trained dogs do, a strange show of politeness; when I touched his head a few times I felt his tension and the hair stiffly risen on his back, like a hyena.

The night seemed much more silent than before; I could hear our steps magnified in this silence and felt exposed, very vulnerable, three generations of my family returning in the dark to our home in Africa.

Even the frogs and crickets had lost their voice. I found mine. To disperse the eerie feeling, I chose to break the rules. I started singing.

It was a song of snow and marching soldiers on desert sands and mountain tops, a song I had learned from my father as a child and that Emanuele had loved years ago, when I sang it to him as a lullaby. Soon he joined me.

> *Lungo le dune del deserto infinito,*
> *lungo le sponde accarezzate dal mar,*
> *oh quante volte insieme a te ho camminato,*
> *senza riposar.**

My mother, now beyond bewilderment, sang with us too. A strange procession.

Finally we reached the spray race at Kuti and the familiar *men-anda* looked ghostly in the moonlight; there were a few impala there. Startled at our appearance and uncertain whether to jump, they stood, fawn coloured and handsome, long-lashed beauties. Then suddenly they leaped high and fled into the night.

We entered the garden from the northern gate, and the other dogs ran to meet us, barking their welcome in a flurry of tails. We were safe.

Paolo, as expected, was reading in a room full of music, drinking his whisky on the rocks, and smoking like every night. When we came in, it took him a few moments to notice that we were covered in mud, still unchanged for dinner, with an air of excitement, like people who have just run a race and reached a goal. We told him the story and he laughed in approval. He would have done the same; he expected no less from me.

---

*'Along the dunes of the infinite desert / along the shores caressed by the sea / oh how often have I marched with you / without a pause of rest.'

'Any encounters?' I told him of the buffalo, of the elephant. Of Gordon's sudden meekness, the silence of the night . . . he raised a brow, did not comment.

Next morning before breakfast he came towards me grinning. He took my hand and guided me out of the garden, along the path we had taken. There, very distinguishable in the drying clay, were our three tracks, my large ones, Emanuele's child feet, my mother's small prints; Gordon's paw marks following close to mine. And a few steps behind us, neat, all along the way, coming from a shrubbery after the middle tank, the rounded pug-prints of two very large lions.

'They followed you all the way, up to our gate.'

The tracks continued alone towards the bush.

Part III

CREATURES FOUND
AND LOST

The Summer of the Crayfish

For my father

Looking on the happy Autumn fields,
And thinking of the days that are no more.

TENNYSON, *The Princess*

O n the opposite bank, two camels browsed with prehensile
lips from the waxy shrubs of rhus. They moved on phlegmatically and in seconds disappeared completely. Harvester ants
busily carried grass seeds, zig-zagging in a scattered procession,
hurrying down their holes. Crouched on my haunches, the sun
hot on the back of my khaki shirt and on my naked legs, I was
fishing.

The brown surface of the water broke in a ripple and a large
muddy claw emerged, to grab the morsel of rotten meat tied to a
string I was holding.

'I got it!' I screamed happily and with a jerk I yanked the
creature out of the water.

There it was, on the murram shore of the dam at Ol Maisor,
a diminutive monster from the underworld, red and glistening,
whiskers sizing up the new environment, legs already tentatively
crawling their way backward towards the dam.

On the other bank, five zebra approached. They stood still a
moment, looked around, and trotted, head down, to drink while
their stallion watched, tail twitching at invisible flies. No breeze
interrupted the dreamy heat-waves lifting from the soil, and from

the thorny treetops came the joyous, powerful choir of cicadas and African birds at noon.

A dust devil swirled over madly and was soon extinguished.

'The largest crayfish I have seen in a very long time,' approved Paolo. 'Amazing how fast they grow here. I thought they only thrived in a cold climate.'

We collected a bucketful in less than two hours and drove back to Ol ari Nyiro in the evening, pursuing a herd of jumping oryx, facing a wild orange sunset, singing an Italian song, and the savannah grass was tall and streaked with mystery in the fading light.

Next day, I drove to Ngobitu's dam and ceremoniously let all the crayfish go.

'One day I'll catch you again,' I let them know. 'You and all the generations you'll produce in all the years to come.' They disappeared into the murky red water, leaving just a few bubbles.

They were early days in Laikipia.

They were days of exploring, of adventure, of love and of bewilderment, when we discovered the new land we had chosen for our home, and ourselves. They were days of youth and fun, when tomorrow was still gilded, when happiness was the morning birds and the evening clouds, turquoise lizards on hot stones, weaver birds building their nests on the fever trees, dung beetles laboriously rolling huge balls, heady perfumes of wild jasmine, and the Pokot women, bare-breasted and in long skin skirts, ambling proudly along the Ol ari Nyiro boundary, head high, minding their cattle.

When I would marvel at buffalo wallowing in the mud baths, at elephant moving silently like shadows through a glade, a few metres away, unaffected by my presence. When I would stand in the night at the bottom of my new garden, almost exactly where the graves now are, a long shawl protecting me from the chill of the eastern breeze, my ears alert to hear the hyena crying on the hills and lion or leopard calling their raucous, rhythmic hunting song.

When my family was still together, and my men with me.

In the months and years which followed so much happened that I forgot my crayfish.

It was the summer of 1981, when Paolo had been a year in his grave and the tree that he had become was bearing its first flowers and the baby he had fathered, but never seen, had begun to talk, that I suddenly remembered.

'Shall we try and see if there are any crayfish in Ngobitu's dam?' I asked Emanuele. 'If they did not all die, there should be lots by now. When was it that we put them in? Five, six years ago?' We could not recall. But Emanuele, a long-limbed teenager with serious-mocking eyes, jumped on his new motorbike, carrying some defrosted meat as a bait, and disappeared in a rusty cloud of dust.

'If they are there I'll catch them. Prepare the dill and mayonnaise, Pep!'

I did not really believe we would find any. No dried-out bleached shell had been discovered along the banks – the remnant of a stork's or heron's meal – to announce their presence. Nor, when we went out for tilapia or black bass, did we ever detect the strange shadows lurking clumsily in the shallows.

So I was immensely surprised when Emanuele appeared, triumphant, an hour or so later, carrying in his hands the largest, healthiest crayfish I had ever seen.

'There are zillions of them there,' he announced. 'We must put them on the menu.'

And so I did.

Simon, my cook, protested weakly, with Turkana pride, at the invasion of weird insects in his kitchen, but finally surrendered when the odd baby lobster, now attractively bright red and crowned with celery and lemon wedges, gained him additional compliments for his already outstanding cuisine.

So it was that it became a feature of hot afternoons on the ranch to go fishing, and no Venetian scampi risotto was ever tastier than the Laikipia crayfish we perfected, with lots of fresh pepper and parsley and a dash of brandy as a final touch.

But it was not just the gourmet satisfaction.

Whenever I see the crayfish at my table, I feel an immediate nostalgia, their appearance and their flavour is an instant joyous memory of other happy days, and I go back in space and season, to the time I was a little girl with pigtails and curious eyes, avidly pursuing excitement and adventure, to the unforgettable Summer of the Crayfish. The first summer I recall with my father.

We lived in the country, in Veneto, in a village called Crespano del Grappa, at the foot of mountain slopes covered in pine, chestnuts and *cornioli*,* where my grandfather owned the silk factory. There my young father, fresh from the Second World War – where he got some medals, lost most of his friends and his dreams, but not his longing for exploring and discovering – practised medicine at the local hospital.

I was an inquisitive child, full of questions, and those early childhood days spent in the country, with any amount of space and the freedom of a large fascinating garden, a wood, an orchard, a vineyard and a stable, were to mark my life forever. As forever would I yearn for the outdoors, the world of imagination, of living things, of leaves and buds, of hidden nests with eggs, of sprouting seeds, of the smell of grass and berries and wild shy creatures.

I was never a town girl, content to be surrounded by the noise and glitters of a modern city. On the pavements, crowded with hurrying strangers, in the lifts, speeding to vertiginous heights, in

* A type of Italian crab-apple.

the shopping malls of marble and fountains, I felt confused and out of place. When, in the years to come, Fate brought me to travel to the world capitals, following the dream and crusade that my destiny had woven for me, I always looked for a window from where I could see the sky; for a place where to walk barefoot; for growing things alive and silent beside the man-made sophisticated junk that stifled me. All this is owed to those early days.

Inside me, bright and keen, there still is a little girl who wants to run, seak the unexplored, who loves the smell of mown grass and campfires, of running water and mountain streams, of low tide and daisies in the sun and cyclamen in marvellous clumps to be discovered in the dense undergrowth.

When my mother went to have her clothes made at the house, outside the village, of a farmer's wife called Ines, who had the gift of magic hands for sewing, I often went too.

Ines worked in her kitchen. It was the largest and most comfortable room in her house, high-ceilinged, cool in the summer, when a breeze moved from the front vine pergola to the shady back garden and the vegetable patch, and cosy in the winter, when the cooking stove was crowded with pots constantly bubbling away their appetizing vapours of hearty soups and sauces.

As a result, to her annoyance, my mother's freshly delivered winter clothes all seemed to exude a smell of minestrone. Weather permitting, they had to be hung out for days before she could wear them, and sprayed with lavender and her favourite perfume, Fleur de Rocaille by Caron, a festive bouquet to me, exotic, inebriating, irresistible.

I was interested in this strange, different house, and observed its details with immense curiosity.

One corner of a large wooden table was covered with a grey blanket and a clean white sheet: on this Ines cut out her patterns, stencilled out of old newspapers, and stitched them with a black Singer sewing-machine.

On the mantelpiece were sepia photographs of old weddings, all the grooms identical with their heavy moustaches and ill-fitting black coats, strangled in starched collars and bow-ties, ears protruding, the only variation the middle or side partings of their brilliantine-sleeked hair. They stood rigid, their hands on a papier mâché column, or on the back of an ornate chair where the brides sat. These were identical, too, jowly, black-eyed, trussed up in corsets which could not disguise their florid good health, stolidly staring at the camera with incurious diffidence.

There were oval portraits of long-lost grandmothers, a faded photograph of Pio X, the *Papa Sarto,* next to vivid depictions of the sacred, bloody heart of Christ, a flame emanating from its middle; a crucifix with a shrivelled olive twig stuck behind it from last Palm Sunday; a doll dressed in pink lace, a few trinkets won at the village fair.

Ines was thin for an Italian woman of her age, with a sallow complexion from lack of sunshine. She wore an apron tucked about with needles, and her permed black hair was streaked with grey.

Chickens sometimes wandered in through the open door, with a blade of sun and dancing midges, and ventured tentatively on the white and black tiles, one foot at a time, observing me sidelong, with red-rimmed eyes.

If the session trailed on too long, I was allowed to go out to explore the sunny fields, provided I stayed within the compound. So it was that one day I ventured to the stream which formed a still pond set in a clump of mulberry trees below the house. I peered into the dark green waters and glimpsed a curious shadow like a long weird insect, hiding below a stone next to my laughing mirrored image.

It was that time of childhood when everything is new and exciting, of vivid physical impressions, smells of new places and animals, tastes of new fruit.

The scent of humid earth, of ferns and cyclamen in the recesses

of the wood's undergrowth; the floury sweetness of the first roast
chestnut, warm and slightly burned; the surprising variations in
the odours of rabbit or goat droppings, chicken coop or geese,
pigsty or stable, of horses and cows; the subtle delicious differ-
ences in the taste of roast pigeon and duck breast; of fresh raspberry or blackberry, currants or small wild strawberry; of milky
new hazelnuts or walnuts, of different types of grapes. The white
nuances of jasmine or wild gardenias, tuberose or lily of the
valley.

It was a constant joy, living in the unpolluted country with a
whole world to explore, and I shall carry with me always memories of 'first times', vivid and fresh with a thousand details.

The intoxicating sharp *corniola* wine, a forbidden sip quickly
offered by a laughing maid, Lidia, a relative of Ines's, when she
took us once to her parents' farmhouse on an afternoon walk, a
blessed diversion which I begged her to repeat. A free new universe unfolded there for me in all its marvels, unknown to my
parents. I helped to harvest, picking with my small hands, the
heavy sun-warm grapes coated in verdigris, and put them in a
wide basket of woven reeds; I watched, mesmerized, the winy
juice spurt purple in the barrels through stamping toes on late
September afternoons; I helped with the milking, sitting in the
dark stable, pungent cow dung heavy in my nostrils; following
Lidia's brothers, I climbed haystacks smelling of freshly mown
grass, lucerne and wildflowers and, astride a cherry tree's branch,
amongst its leaves like a bird, I ate cherries until I felt like burst-
ing; copying the other children, I kept the stones in my cheek
and spat them out all at once, trying to hit my friends and laughing with abandon.

Small things were treasures, mysteries, objects of curiosity and
investigation: a dead water rat in a mossy hole, with stiff whiskers
and bared yellow teeth, a lizard decomposing in the sun by a
clump of forget-me-not; drinking reverently from the sacred
spring in a shady glade below the hill of Covolo. There – Lidia

narrated – the Virgin Mary appeared to a lost, pious child shep-
herdess, and stuck three of her holy burning fingers into the rock,
creating three holes from which fresh clean water spurted
instantly. Like a minor Lourdes, the springs of *Tre-Busi** had
become a place of worship, and a monastery had been built up
the hill, destination of many of our afternoon walks. We reached
it along a steep, dusty white gravel road, flanked at regular intervals by *capitelli,*† filled with votive tributes of flowers and silver
hearts, where the story was told in naïve oleographic frescoes.

> *Oh donna donna*
> *donna lombarda*
> *vieni stasera a ballare con me.*
> *E io si*
> *che vegnaria*
> *ma ho il marito e non e non posso venir . . .‡*

We returned singing old ballads, of love stories and betrayals
and honour vindicated, and passion and tears, and stopped half-
way for our *merenda*§ of bread and cheese and raspberry syrup,
sitting on the grass in the shade of a chestnut tree. There, looking
down at the valley, I loved to search out my grandfather's house,
disguised by a green clump of Cedars of Lebanon, next to the
filanda§§ which was always easy to find, marked by its tall, smoky
chimney made of old bricks.

In this bucolic contest of magical country delights, the discovery of the quaint, odd fish was intoxicating.

That evening I could not resist the temptation to recount it,
and described to my father, with a wealth of details, the mysterious creature that I thought I had seen. My father listened, with

---

* 'Three holes' in Venetian dialect.
† Shrines.
‡ 'Oh you woman from Lombardia/come tonight and dance with me./Oh yes oh yes, I would
come with thee/but I am married and cannot go.'
§ Snack.
§§ Silk factory.

the attention he always paid to my stories, and understood; and thus began the magic, exhilarating Summer of the Crayfish.

My father, with the enthusiasm and abandon that destined him to be forever young, forever inspiring, loved to find special gifts of wild morsels, gourmet foods that nature reserved for her followers, the tireless walkers, the early risers, the mountain climbers, the ones who deserved the privilege. He told us stories of handfuls of raspberries picked while running from enemy fire, of *porcini* mushrooms discovered on perilous banks inaccessible to the lazy; and the crayfish discovery coloured those mid-summer days with excitement and anticipation.

'They must be crayfish, so rare in these waters, and quite delicious. We'll go fishing tomorrow night, and you will come, too, because it is your discovery and you must guide me.' My father knew how to touch a deep chord of pride in my eager small heart. I was perhaps five, and not normally allowed out after dark: the exception lent my day an unbearable thrill.

We left together in the evening like conspirators, taking with us a packed dinner of sandwiches, torches and blankets and bits of meat and old salami as bait. I brought my father to the pond in the clump of trees. We peered, searching the water with the torch. My heart beat wildly: had it been an illusion? On the gravel of the bottom, amongst the stones and watercress, two grey shapes armed with claws were fighting silently.

'Well done!' approved my father, my hero, and the night glowed with my tumultuous pride.

The imprudent shellfish, attracted by the meat effluvia dissolving in the current and by the torchlight, were easily captured, and put into a tin bucket. I helped, holding the torch, carrying the bucket, and my father taught me how to grab the crayfish, just below their pincers, so they could not nip us. Splattered in mud, wet and shivering, I fell asleep exhausted in my blanket on the way back.

The crayfish were quite elusive. We could only catch a few at the time, not enough for a meal. We decided to store them at home, in one of the tanks at the back of the silk factory, a deep old cement tub where it was exciting to see the ghostly figures disappear in the mossy depths.

We went back almost every night. My father pointed out remote stars and taught me their names, we sang the partisan songs of the last war, and I was elated. The flavour of escapade, the new independence, the sounds of the night appealed to my imagination, and the experience had borders of reverie like a summer tale.

As the season progressed, the shapes became numerous in the silk factory tank.

It was finally September, the time of harvest and of returning to town, where, for the first time, I would go to school. The night of the crayfish dinner arrived.

My father surprised me by not helping with the recapture of the crayfish from their tank: this occasion was tame, and a net was produced.

He only appeared at dinner time, when already a delicious aroma was filling the house, of white wine and olive oil, and herbs and garlic.

I had been worried about eating the creatures I had come to know so well, but when they appeared on the silver platters, crowned with lemon wedges, I was startled to see my old grey friends changed to red glistening jewels. Their transformation took away my guilt. They no longer looked like those little monsters I had observed fighting in the still, silent waters and, following the others, I ate them with glee. Of the flavour, I only remember it was one of the most surprisingly delicious I had ever tasted.

No one knew that secretly, that afternoon, I had taken two couples from the full buckets and put them back furtively in the tank where they had survived so well for a month. By now, a thousand crayfish perhaps fill the tanks and ponds of that forgotten garden.

That evening, while the lamp threw weird patterns onto the white linen table-cloth monogrammed with my grandmother's initials, I looked at my father across the table, pensively sucking a claw, and knew that he already missed the adventure, and would have approved of my secret.

And once more, in his green, short-sighted eyes, I recognized myself.

Remember the Seagulls

For Livio

But Jonathan Livingston Seagull was not a common bird
... more than anything he loved to fly.
RICHARD BACH, *Jonathan Livingston Seagull: A Story*

During one of my rare visits to Italy a couple of years ago, returning from a delicious lunch of fish and chilled wine in the *valle** of a friend, I found myself driving through the piazza of Jesolo which had been so familiar to me over twenty years earlier, before I came to live in Africa and turned that page of my life forever.

There was a grey stillness in the early April afternoon, a familiar humidity, and the smell of freshwater seaweeds from the invisible lagoon came through the car window, bringing back memories, and a sudden nostalgia.

On the spur of the moment I turned to my friend Marisa, companion of many expeditions and European peregrinations.

'Shall we go to Cavallino? I haven't been back since 1972. It's only a few miles from here.'

She turned her red head and the gold-flecked eyes flashed the quick smile which had earned her the nickname 'Fox'.

'Why not? Are you sure you can handle it? It will be changed. Maybe no one is there.' She had been one of our frequent guests in the past.

*A geographical characteristic of the Venetian lagoon.

116

'Let's try. It will be ages before I come this way again. Perhaps I never will.'

With instant silent agreement she steered her silver-grey Mercedes to the right, and in a few moments we crossed the small bridge, out onto the white gravel of the dirt-road on the canal dike, and we were on our way.

The slopes were covered in fresh tall grass; a few primroses had begun to sprout, and creamy calla lilies along the banks on the water edge.

I recognized the turns of the track, the narrow, late-nineteenth-century red-brick peasant dwellings on the other side of the canal, small ploughed vegetable patches, a vine twisting into a pergola, fields of asparagus, a row of peach trees. Somehow things looked smaller and the distance was shorter than I recalled.

The road turned sharply right, the view opening out to the sea, the island of Burano in the distance, and there, suddenly, was a courtyard, a gracious brick house covered in green creepers mirrored in the still water of the lagoon, familiar outbuildings. We were there.

Our car came to a halt, skidding on the gravel. There was no other vehicle parked in the drive, and the dark green wooden shutters were closed. It looked as if there was nobody around. After a moment's hesitation, I opened the car door.

A muffled noise of voices and boxes dragged on a cement floor came from a low construction on the right, below which there was a boathouse built straight on to the darkish water. Framed through its porch were a few white egrets in the distance, standing amongst the rushes on the *barena*,* which looked as barren and wild as ever. Two large seagulls flew over slowly, and as I followed them with my eyes, I noticed that someone was looking out of the *cavana*.† I turned.

There was no reason why he should have known me after

---

*A sandbank or shoal.
†Venetian term for boathouse.

twenty-two years, appearing unexpectedly, out of context as I had, but I would have recognized his face anywhere.

He wore waders and an oilcloth windbreaker over a sailor's sweater as he had the last time I had seen him. The woollen cap covered half his face, and kind black eyes lit up amongst the wrinkles dug by years of squinting through autumn fogs, of peering into shallow waters for darting sea-bass spawn. His skin was burned a dark copper even though he had sheltered it with calloused hands while scanning the sky in countless summer suns, trying to spot the first duck migrating back from the north.

It took him only a moment, and all at once his arms lifted in surprise and welcome, his hand was raised to his cap, his mouth opened in the widest grin, and, without taking his eyes from my face a moment:

'*Maria, Signor, Gesù varda chi che ghe xe!*' he exclaimed in Venetian dialect. '*La signora la xe tornada. La signora Kuki!*'* We moved forward at the same time and almost embraced. We halted, holding each other's hands.

'*Livio! Nol xe cambià. Come falo a recordarse de mi?*'† I answered automatically in an idiom I had not used in years. I was overwhelmed. Although I had hoped that I would find him, I had not really expected to do so.

Livio had been Paolo's headman in the *valle*, the head fisherman, the head hunter and the game warden. A typical product of the *valle* where he was born and whose secrets he knew, he was the invaluable collaborator on whom Paolo had relied totally. Competent, honest, of good disposition, he loved his job and excelled at it. Over the years he had become a trusted friend.

Then Paolo went to Africa and never came back. When Paolo died I received a cable: 'We send you and the children our heartfelt condolences for Dr Paolo's sad demise. Famiglia Dalla Mora.'

I had not had contact since.

*'Mary, Lord, Jesus, look who is here! Signora has come back!'
†'Livio! You haven't changed. How can you still recognize me?'

With instant recognition, he greeted Marisa, too; she had not changed much since the old days.

I wondered at the long memories of people who live in remote spots. It happens always in Africa: one meets someone in some god-forsaken place, and is recognized instantly, as if nothing had happened to fog the memory in the intervening time.

He insisted on giving us a *caffè corretto,** and showing us the house. With trepidation I entered rooms that had been familiar, climbed stairs that had felt Paolo's rapid step, Emanuele's little boy's run.

I looked out at the unchanged expanse of water from the small windows of the altered dining room, where the sound of laughter and popping corks had risen high in days gone by. Only the tall smoky chimney had not been touched.

Cavallino had been Paolo's and his first wife's place. I had never really considered it my home in the three years or so before coming to Africa, when I had often stayed there, while I was still living with my mother in the country. Yet I had felt its spell, the charm of its old stones and sleepy past, the beauty of the lagoon bathed in the mercury and coral of September sunsets, the ancient rhythm of traditions on which the life of a fishing *valle* was based. For Paolo, it had been a consolation while he waited to return to Kenya, the place to which he had discovered he belonged.

We went from room to room and sat finally, sipping our coffee in the dark dining room, where game wardens and *valle*-hands had gathered after long fishing or shooting mornings, to eat their earthy meals of crab soup and grilled eels with *polenta,* and roasted chestnuts washed down by bottles and bottles of new wine.

When we went out I noticed that a lone seagull was flying high above us and Livio followed my eyes.

*Coffee with a measure of strong liqueur added, traditional in Veneto in winter.

'Do you remember,' he smiled at me, 'the time I gave you the seagull chicks?'

And memories flew back on still wings.

One afternoon, in the uncertain hour of opal before night falls, when the lagoon looks like a fluid mirror, Livio had returned from his daily tour of the locks and docked the boat inside its shelter.

Below his dark woollen beret the weather-beaten fisherman face smiled, and a hundred happy wrinkles marked his red cheeks.

'I have a present for you,' he offered in dialect. 'Come and see.'

On the bottom of the boat, beside a coil of rope on the tar-blackened planks, two brownish, fluffy birds crouched, from whose open beaks came desperate screams of hunger and terror.

'They are *magoghi*,* said Livio proudly, 'just hatched. There were still broken shell pieces stuck to their plumes and all over the nest.'

Perhaps I did not manage to thank him. I felt bewildered and guilty that my love of animals, which was well known, had brought about this unconscious cruelty.

Through my mind darted plans to return the miserable new-born creatures to their marsh, to the beautiful destiny of which they had been forever robbed: but the sun had set, dissolving into the water in reflections of red and mother of pearl, and there was nothing to do but agree to look after them.

It was a time when I wore long skirts in bright cotton, partly to protect my legs from the damp without wearing trousers which rubbed against the still fresh scars of my femur fracture, partly because I felt they matched the newly found peace, the bucolic pleasures and simplicity of country life in the Venetian *valle*.

*Young seagulls in Venetian dialect.

I gathered my skirt into a nest and cautiously I laid there the birds. The gaping beaks with the pointed pink tongues, the greedy red-rimmed eyes and the ungainly cackling were far removed from the majestic grace of the adult royal seagulls they would one day become.

With little originality I remembered the story of the Ugly Duckling which like many fairy tales is pervaded by the revenge of the small and humble against the wealthy and handsome. Cinderella, Snow White and armies of Thumbelinas had their animal equivalent in the ugly cygnet vindicated by Hans Christian Andersen. So, feeling like a fairy godmother, I set about caring for my wards.

To begin with I looked for an appropriate container. An old dolls' pram in green plastic looked perfect, and I lined it with wood shavings, small pieces of materials and clean rags.

Emanuele captured small fry from along the shore and presented them to me.

While I prepared their crib I kept the baby seagulls tucked into my skirt, working with one hand without noticing that the chirps had grown weak in the warm darkness of the folds. When I tried to put them in the pram they were asleep. I contemplated them for a while, still in my skirt, soft tiny balls of feathers like a Victorian muffler, one with its head below the other's wing.

When hunger woke them up hours later I was numb with stiffness.

I quietened their shrieks of outrage, forcing the fish that Emanuele handed to me down their wide throats. They devoured it voraciously, with ease born of instinct, pecking my fingers in their greed. That was probably their first meal, and it was thrilling that they accepted me so quickly. When there was no more fish, I covered the pram with a rag, and in the darkness they slept again.

That night I read what I could find about aquatic birds in

Paolo's extensive library. So I came to learn of a curious experiment, performed on baby ducklings.

When the eggs were about to hatch, a few were removed from the nest and put into an incubator, a surrogate nest, where, instead of the mother duck, there was a rectangular cardboard box, tied with a string so that it could be pulled along.

The moment they broke out of their shell, the ducklings went immediately to the cardboard shape, nestling under it and moving with it whenever the string jerked it away.

At that stage they returned the ducklings to their real nest, where in the meantime the others had hatched, and were gathered around the mother duck, following her every movement. The experiment-ducklings did not join them. They remained confused, cowering in a corner until the box was placed near them, and they rushed to it with jubilant chirps. Thus was the theory confirmed that birds recognize and are imprinted by the first moving object they see after hatching from their egg.

When I put them on the bank and retreated a few paces to look at them, they rushed towards me with piercing screams, relaxing only when they managed to huddle under my skirt.

I experimented several times, my heart in my mouth, and the result was always the same. They did not follow anyone else, on the contrary they shrank from people, but at my appearance their cries changed as if they definitely recognized me. I realized that something similar to the experiment I had read about had happened to the wretched young seagulls; they had chosen me as their mother substitute.

The revelation that two royal seagull chicks, among the noblest of birds, unsurpassed masters of sea and ocean, dancers of the sky, had chosen me as their mother, humbled me, made me proud and scared me with the responsibility. In my role as mother-seagull I felt awkward and inadequate.

I decided that I must try, at least partially, to fulfil my task by teaching them to swim.

It was May. Already the days were getting longer and the sun warmed the chill water of the *colauro** where captive sea bass fattened, waiting for autumn, and eels the colour of mud slid in the weedy bottom with disquieting movements.

Next day I gently set the seagulls on the grassy slope of the canal, where seaweed and bleached crab legs dried in the sun. I moved with tentative steps into the oily cool water, feeling the mud squirm beneath my toes when they disturbed a flatfish or an eel.

The gulls came to the water edge they did not yet know, and there they stood, uncertain, stumping in the mud with spread claws, crying.

I moved deeper into the water, imitating their sounds like a farmer's wife when she calls her chickens, a comical scene to the uninitiated. But the only spectator was Emanuele, connoisseur and lover of animals and loyal supporter, who was following events from the shore, biting his finger with concentration. He was no more than four years old, and it must have been an extraordinary adventure.

Step by cautious step, shrieking unceasingly, the seagulls followed. Then they launched themselves, leaving behind them a touchingly tiny wake, like frail paper boats obliquely drifting in the current. They swam towards me with increasing confidence. From the shore Emanuele laughed and clapped.

Finally soothed, in their element, with an instinctive fluid elegance which their former awkward steps had not foreshadowed, the *magoghi* floated around my head, their tiny eyes for the first time at the same level as mine.

Every day after that first attempt, we swam together and I fed them the small fish that Emanuele managed to capture for them. Ten days or so went by.

The time of the unavoidable first separation found me nervous

*Venetian dialect: part of the canal that links the *valle* to the lagoon.

and uncertain, like a real mother about to leave her own newborn behind in the care of strangers. A certain premonition made me fear that something would happen to them in my absence.

It was the dog which in the night after I left broke into the bathroom where, after having fed them more than usual, I had confined them to keep them safe and comfortable.

It was a pain that I had never experienced before, a sense of failure, as if I had betrayed them, failing the universal rule of caring for the young.

It was a lesson reinforced when later I went to live in Africa, never to encourage the gift of a wild creature, which captivity would kill or make into a slave, a worse fate still.

And I saw a new meaning in the large seagull which followed our attempt at swimming, soaring and gliding in large circles over our heads, calling raucously in vain.

I looked at Livio. The afternoon was drawing on, it was getting late. The grey fog of every evening, with its drifting ghosts, was rising from the water. Time to return.

'The seagulls. Yes, of course. It was sad when they were killed, but perhaps it was as well. As much as I loved them, I could never have taught them how to fly.'

We embraced with promises of writing, keeping in touch. I knew I might never return.

We drove off. I looked back before turning. The house was fading in the mist, silent, with all its past untold and unrevealed.

Livio's figure was silhouetted against the sky, a hand still raised in farewell.

The seagull had disappeared.

The Brigadier's Cheetah

The sleek and shining creatures of the chase
TENNYSON, 'The Revenge', v, 147

Sometimes in the afternoons of the holidays, when we were in Nairobi, at our house near Rosslyn, Emanuele asked me to go with him to see Tigger: he remembered the night when we had encountered him, and how he had not run away when he saw us.

Tigger was a male cheetah, a few years old, who lived on the coffee farm of one of our neighbours, a retired army Brigadier.

We used to jump into the car, and set off, leaving the tarmac road for a red murram track winding through coffee bushes. Soon the usual scene would appear. Along the slope they walked slowly, among the tall dry grasses, their patient black dogs following quietly, wagging their tails in rhythm. A girl pushed a pram from which peered the freckled faces of two children. Next, tall and slightly bent, leaning on a golf club, came the Brigadier, with his wife, and the cheetah.

The cheetah moved lightly, the fluffy tip of his tail barely brushing the ground. His easy gait had the grace of a dance rhythmed to the silent beat of distant drums. A leash circled his chest. He advanced sure and languid, his small head buried between powerful shoulders, his one eye yellow and alert, its colour matching his coat maculated with regular black spots. Emanuele ran out to hug him, and the rough tongue licked friendly his young neck and cheeks. The eye closed in pleasure, and the cheetah purred like a large happy cat.

This was the evening walk, amongst the coffee bushes along the slopes behind the house, the same now for years. Ever since he had been found, with two other cubs, after their mother had been killed during a long chase in the savannah, he had been kept and cared for, chosen from among the litter because of a congenital eye defect that would never allow him to hunt alone and to be independent like his brothers.

When they called him 'Tigger' he was just a kitten, as soft as a real toy and defenceless as any rejected puppy. They called him Tigger, a fierce name, but he was gentle. Only when wild rabbits darted from their holes in the red earth to disappear in mad leaps through the undergrowth, did his young muscles tense, and he would coil in the ancestral instinct, ready to spring after a running prey.

For play companions Tigger had three Labrador puppies, recently born in their mother's basket behind the outside staircase, which the Golden Shower covered with cascades of orange blossoms, sheltering it like a real lair in the forest. The playful fights with the puppies on the lawn, the habit of sharing bones, the sudden frantic sprints and the deep sleeps of abandon, tummies up on the grass, and eyes closed against the implacable glare of the Equator's sun, made them brothers and inseparable. He was never alone, and who knows if in the flat compact head dreams ever drifted of runs over plains in the short red sunsets, when the Highlands plains are alive with herds of gazelle, and predators emerge from the shadows of the day with silent steps, to sniff the scent of the preys of the night.

Often, at a sudden movement, he lifted his head. His round black nostrils vibrated sensitively to imperceptible scents, and the ears tensed to the inaudible shuffle of near, secret lives. The amber-and-honey eye scrutinized the horizon like an eye-glass:

even the faintest shiver in the savannah grass did not escape his gaze. A black line circled his eye sockets like a mask and defined his round features, dividing them with two black tears which slid down to the corners of his mouth. In his black war paint his face looked sad.

He played with the dogs, more dog than cat, lacking retractable claws; he looked like a large greyhound, to which a feline head had been attached. His fine legs were far slenderer than the powerful, stocky ones of a lion; his paws far slighter than the rounded catty ones of a leopard. He moved lightly, with a lazy elegance born of the assurance of being safe.

His mother had given birth to her cubs alone, in the shade of the large acacia, amongst the low shrubs which protected her from the eyes of her natural enemies, the hyena and the wild dogs. Alone she had brought them up, leaving them for entire days while she stalked the plains, alert and sinewy, her graceful tail flowing like a mane in the winds of adventure.

They had soon learned to be independent, helped by nature which had given them a crest of odd long hairs on the top of their heads, white and straight to match the long highland grasses bleached by the African sun. This allowed them to hide, camouflaged even amongst the driest of sticks and thorns, until their mother returned at dusk, to feed them with swollen breasts, the smell of a Thomson's gazelle's blood still in her breath.

But one night she did not come back, and the hunter found them, hungry and torpid next day, huddled together for comfort in the shade of the acacia tree.

He grew with the puppies and he looked funny and out of place among the black coats; his white tuft stood out pathetically on the green lawn, so different from the silvery waves of the savannah grass stirred by sudden winds. He grew swiftly, unaware of his great strength and grace, until he was finally an adult: and in the nights of March Tigger suddenly felt the call of his race, even though the farm where he lived was on the edge of the city, and the noise of passing cars and lorries filtered through the narrow cluster of forest like an intrusion into his loneliness.

A tame female cheetah lived not far away, in a garden protected by bougainvillaea, dogs and high hibiscus hedges.

One night we came back late from a party, and, at the corner of our road, amongst sisal and giant poinsettias, Emanuele's sharp eyes distinguished a motionless shape, outlined by the full moonlight in all its wild beauty.

He sat perfectly still, like a statue of a sphinx, under the pepper tree, next to the sign board inscribed with our name. His neck was tense, his nostrils flared to sense the breeze. A low, breathy noise, a deep intermittent growl, perhaps a mating call, came from his splendid throat.

The dogs in the neighbourhood began to bark furiously from all the gardens, howling their unrest to the moon. Neither this commotion nor our approaching car seemed to disturb him.

'Tigger,' whispered Emanuele softly from the car window.

Tigger turned his head slowly, to look straight at us, fearless, remote, surrounded by his mystery.

A moment, and he was gone, swallowed by the darkness.

It was after then, I imagine, that the legend of the leopard of Rosslyn was born. Someone else saw him, and gave out an imprecise description. Everybody locked in their dogs at night for a time, as the leopard's predilection for eating dogs is well known.

Not us. We knew it was simply Tigger who wandered for miles and miles, from his safe dog's bed and warm blanket to answer the call of his dormant, never fulfilled instinct. We had phoned the Brigadier, and he saw that the gate was still locked but the basket was empty in the moonlit night. Even if there was no need, as a cheetah can jump high and swift, he left the door of the enclosure open.

Next morning Tigger was there, impassive and tame, as if nothing had happened, to play with the dogs, and to wait for the leash and his evening walk among the coffee bushes: like in an old print, sunlit from the back.

Some time later, to everyone's wonder, our neighbour's female cheetah gave birth.

One of the cubs had an inherited eye defect, and we all knew what had happened.

The Tale of Two Bushbabies

And the elves also,
Whose little eyes glow
Like sparks of fire, befriend thee.

> ROBERT HERRICK,
> *The Night Piece: To Julia*

Nights in Africa are never silent. If you listen carefully an entire orchestra of diverse sounds and secret voices reaches you from the grass and the hills, the dunes, the ponds and the trees. And if you look for the unseen creatures which animate the night you can often, for a moment, glimpse their eyes, piercing the blackness. If they seem to dance high upon the treetops like mischievous elves, faster than your sight can catch them, and if their voices sound like the whine of a child lost in the forest, probably they are bushbabies. Related to lemurs and monkeys, nocturnal, arboreal, they feed on insects and fruit.

The first I met was Koba. From the dark of the store you could only see his eyes: shiny and round, and in some odd way disturbing. Opening the door, I brought the sun with me, and in its white blinding light the pupils of his huge eyes contracted, and the irises stood out dull in the tiny face.

He was clasping the shoulder of the man who wanted to sell him; a string of plaited palm leaves circled his narrow waist and underlined the difference between the upper part of his body – slim, with frail arms – and the lower half, with powerful, muscled legs, ready to jump. He looked like a small kangaroo with a squirrel's tail.

I had never seen a bushbaby so close before. The nights at the coast are full of their raucous screams, as they jump from branch to branch on the baobabs. But during the day they are invisible, and only after sunset is it possible, with much patience, to discover their lithe bodies amongst the leaves.

It was easy to catch them.

The local people in those days hung from the baobab branches dishes of a strong sweet beer, made out of coconut and honey, and the bushbabies could not resist its heady aroma. Dead drunk in the green dawn, still in a stupefied sleep, they would lie scattered at the foot of the majestic trees, like the moths they had not managed to eat. The villagers would harvest them easily, and they would try to sell them, tied to a length of palm twine, to passing tourists at the ferry-boat jetties along the creeks.

Koba had been captured by chance. He was still a baby, and had been grasping his dozy mother's back when she had been caught. But she had managed to escape somehow, and the man, who was the swimming pool attendant at the hotel in Diani, had brought the baby to the pool store, where during the day he was kept amongst bath towels, deck chairs and rubber flippers. It was there that Emanuele had discovered him and, in dismay, had begged me to buy him and let him go.

Now the young bushbaby stood trembling, uncertain and fragile like a bird who cannot yet fly. He looked strange, somehow alien.

Overcoming an instinctive, inexplicable repugnance, I reached out to take him. His small black hands were damp and gluey, with the wrinkled knuckles and spatulate nails of an old child; they were cold in the great sun. He jumped to my shoulder and grabbed my neck, and against my hair I sensed the pulse of his small frightened life, seeking protection. I suddenly felt for him, and for a few shillings he became ours.

We called him Koba, a contraction of the Swahili name Komba, and Emanuele and my young stepdaughters were deliriously happy of his gift. My condition was that in time we should let him go.

We tried. We had learnt that initially his mother had come to look for him many times, piercing the starry night with shrieks of pain, leaping from the palm trees onto the flowering shrubs round his prison. Emanuele put him out on a branch at dusk, and waited several evenings for her to return: but she never did, nor did Koba dare to go off on his own either.

Finally we decided to keep him with us.

When the holiday came to an end, we had to leave the coast and return to Nairobi, but there were bushbabies there, and we hoped that one day we could let Koba go free.

He was a delightful pet. He slept curled in a ball in some hidden recess, on a bookshelf or above a window, and woke up to eat the fruit or the insects which the children constantly caught for him. He often came to us hoping for a special morsel, a sip of the sweet warm tea that he loved, his head slightly tilted as if waiting, his round velvety eyes immense, attentive and oddly unblinking. He took our gifts with dainty fingers, and ate them slowly, holding them with both hands like biscuits.

There was something curiously disquieting about him, and I could never fix my eyes on his without unease, as if he were the memory of a lost identity, lingering in the subconscious, like the faded image of an unknown prehistoric ancestor.

Apart from the repellent habit of urinating on his hands and leaving a trail of humid scent which smelt strongly of liquorice and overripe papaya, Koba was quiet and gentle. But our dogs were unsettled by his presence, pricked their ears at him, and growled in warning – or was it perhaps jealousy at our attentions – whenever they saw him grasping a curtain and repairing to a high shelf inside the house. We realized that we could not leave him at large on his own and, much as I hated it, we had to build a vast cage for him, enclosing enough leafy branches so that he could jump about safely whenever we went out.

One day I came back to find the cage door open. The leaves seemed disturbed and Koba was no longer there. The children called him that evening, and the evening after, and many more. Emanuele put out ripe mangoes, passion fruit, fat glow-worms and grasshoppers on the forks of branches to tempt him. But Koba never came back. To console Emanuele I told him that Koba was probably happy and had found a companion. Yet I knew that he could not possibly cope on his own.

It was an afternoon a week or so later, and it had been raining. Leaves covered the grass below the forest trees in my garden at Gigiri. Termites on their nuptial flights filled the air with their rich, merry buzz, and I inhaled deeply the humid air which smelt of rich earth, humus and growing new shoots, while I walked around my garden with a friend who had come for tea.

It looked like a discarded *peluche,* a toy of a lost childhood forgotten out in the rain, a kitten drowned in a Venetian canal, its wet hair sticking pitifully to a miserable, minute body. It lay next to the tree which it had not managed to climb. My friend gasped in horror.

Koba's tail seemed made out of damp feathers, but what haunted me for nights were his eyes: open and glassy, totally white, with no more pupils, they were like winter mirrors that had been dulled, and reflected no more.

We buried him quickly behind the staff quarters, so that Emanuele would not know.

The children continued to call him in the evenings, from time to time, and on a few occasions they thought they saw him leaping in the treetops like a fairy. I never dared to tell them what had really happened. We all missed him. The memory remained of the faint smell of liquorice and overripe fruit which was his, and a sadness like a sense of guilt.

And for a time I could not hug my large dogs, although I knew that they were not to blame, really.

Years later, one afternoon in Laikipia, a plane landed at Kuti and a couple of friends strolled over to the house. Something seemed to move under the girl's jacket. Round eyes, unblinking, peered at me from its folds and brought back the memories.

'It is a young bushbaby,' said Davina. 'Could you please keep him: the others have rejected him; they wounded him in a fight.'

Davina's mother had a family of semi-tame bushbabies living in and around her house in Karen.

'He is too young, you know. They are territorial and they are going to kill him.' She gazed up at my tall roof of *makuti*.

'He will just love it here; please take him.'

I looked at all my dogs. The image of Koba's white eyes chilled me for a moment. The bushbaby grasped my hand with tiny sticky fingers.

'Why not? Provided you accept that I shall never put him in a cage.' I shivered at the thought. 'Bushbabies should be free to escape. He may find a friend.'

I knew we had bushbabies in Laikipia, particularly in the Enghelesha forest, although I had never seen one yet.

I called him Charlie. He spent the days sleeping in the tiny wooden bird house the *fundi* built for him, and which I hung from the tallest post on my verandah. In the evenings, when the bats filled the air with their screeches, he woke up and went to feed from a bowl of fruit and cake crumbs and honey. Then he crept out to the windows, trying to catch the odd night-insect attracted by the spotlights reflected in the glass.

He loved my ceiling. The *poriti* rafters made from mangrove poles, and the palm fronds of the *makuti* roof may have still smelt of the coast from which his ancestors came, and provided him with the most exceptional palaestra for his jumps, mad exercises and somersaults. Every night he performed there for us, pirouetting wildly above our heads in daring leaps like a circus acrobat.

During dinner he used to approach the dining room table in expectation of some special gift of a choice tid-bit. He loved chocolate soufflé, at which Simon, my Turkana cook, excelled. And when he could smell hot vanilla, he waited for me to stand calling him softly, holding out to him a piece which he ate politely, looking at me intently, with his thoughtful round eyes.

One should really never ever keep wild pets in a tame home. It is impossible to give them the constant care and attention they need. They develop habits that they should not have, and become dependent and over-trusting. Usually it all ends in tears.

I went to Europe for ten days, and when I came back Charlie was no longer there. In the evenings, during my absence, there had been none of the lights to which he was used, none of the animation, the activity, or the human life, unfolding below in front of his curious eyes.

Bushbabies are gregarious and perhaps, as he had felt lonely, he had wandered off. An eagle owl, that eats kittens, squirrels and rats, had been seen often, flapping its heavy wings in the moonlit garden, its hooded eyes scanning the flowered bushes for a sudden stir. The first night I came back I spotted its large sinister shadow perched on the yellow fever tree in the middle of my lawn, and its raucous scream of hunger before the hunt sent a cold shiver down my spine.

I felt sure that he had taken Charlie. I was sad, and so was Emanuele, then a teenager with long legs and wise eyes, and also Sveva, a chubby toddler who had adored watching Charlie's antics, and giggled at his passion for vanilla and chocolate desserts.

It was perhaps a year or so later that Rocky Francombe, the manager's wife, whose house at Centre was about eight kilometres away from mine, told me with great excitement:

'We saw a bushbaby last night. He came over to the house while we were having our pudding, and climbed onto the verandah beams as if waiting for us to notice him. He ate some of the passion fruit mousse from my hands. Today Andrew found him asleep in a bougainvillaea bush, next to our bird bath. He had taken over an abandoned starling's nest. I think he is Charlie.'

She smiled at me.

'He was not alone. There was another bushbaby asleep with him.'

Emanuele's Chameleons

On a souvent besoin d'un plus petit que soi.*

JEAN DE LA FONTAINE, *Fables*, II.11

'Le Lion et le Rat'

I remember him well,' said the pretty young woman I had just met, looking at me with a shy smile. 'We were in the same class at school, when we were children. He was kind, quiet and different. I was sad when he died.'

Her eyes in the darkness seemed misty – or was it a candlelight illusion? 'He always kept chameleons in his desk.'

'Pep, look what I found!' A greyish miniature dragon clung tenaciously to his straight blond hair. I gasped. It was an ugly thing, with rough skin covered in round dry blisters, three crested protuberances on its nose, curiously like a rhino, and a large toothless mouth, frog-like and quite repulsive. With consummate gentleness Emanuele disentangled the creature, and held it out for me to see.

It was a March afternoon in Nairobi, after one of the first sudden showers of the long rains, which leave an intense smell of wet soil and fresh hay and are followed by a violent sun that instantly dries the drops trapped in the grass. Emanuele looked at me with his deep eyes of brown velvet, shadowed by unknown melancholies older than his years.

*'One often needs someone smaller than oneself.'

137

'It is a Jackson's chameleon, Pep,' he said proudly. 'I found it in the bamboos.' He contemplated it admiringly. 'Don't you think it looks like a *Triceratops*? Can I keep it, please? His name is King Alfred.'

King Alfred was the British king who fought against the Danes. In Emanuele's history book they wore the horned helmets of the legendary Vikings. I suppose that first chameleon owed his regal name to this association.

I nodded weakly. A quick glint of triumph lit his eyes.

So began Emanuele's love of reptiles, his passion for chameleons, and an extraordinary capacity for finding them wherever he went. He had been born a collector. As a small child he had collected minerals, shells and small model animals. Later, snakes would take over as his abiding passion. When chameleons became the first reptiles he officially owned, I did not know that we had entered a new era, and that there would be no going back. He was six. In a few years snakes would lead him inexorably to his destiny.

Soon we all liked King Alfred.

He did indeed look like a gigantic herbivorous dinosaur which had existed in the Cretaceous period, as I learnt from one of Emanuele's books which I went to check for reference. In fact, dinosaurs had entered our household years earlier, when my father, during his peregrinations, had discovered a sensational deposit of their bones on a fossil river bed in the Ténéré desert. A black and white photograph of Emanuele aged four, reaching out to the tall skeleton of a monstrous *Diplodocus,* to impress the reader with its size in relation to our miserable human proportions, appeared in a book in which my father wrote his adventure.

Unlike me, Emanuele knew all about dinosaurs, their looks, names and habits, and undoubtedly there was a very strong similarity between the original *Triceratops* and its enigmatic descendant which had come to live with us. A chameleon is a creature of marked individuality, and I could easily see how a curious and intelligent boy, fascinated by animals, could be mesmerized by this deliberate and friendly little monster.

During the day, King Alfred lived in a box full of leaves and small branches. Emanuele fed him with insects which he captured at school, whenever he had a free moment, and kept in an old jam jar. Often, however, he smuggled the chameleon to school in a small perforated cardboard box, and at break time would let him climb on the low shrubs in the school yard, and observe, enraptured, the antics of his hunt.

King Alfred was a few inches long. His legs ended in hands shaped like pincers, with strong fingers sprouting at opposite angles to allow a sturdy hold on the flimsiest of stems and shoots. His curly and prehensile tail could twist quickly around the most minute asperities of leaves and twigs, with a precarious sense of balance, like a monkey diving from the highest trees in the forest.

His most striking feature, however, was his eyes, stereoscopic instruments independently rotating to focus, through the holes of the irises, along a narrow field of vision that ensured the infallible aim of the spring of his viscous tongue.

An unsuspecting grasshopper swayed on a blade of grass; his tongue would dart out faster than our sudden repugnance, which lingered well after the insect had disappeared into the cavernous mouth. We caught our breath in horror.

In time, however, I grew used to this revolting performance and even found a certain fascination in watching its precision, which reminded me of the skill of a cowboy's lasso, or of the cruel, mindless catapult that interrupts the free flight of a bird.

It was the strangest thing of all to watch the colours vary in King Alfred's grainy skin. Brown in the sun, he changed to unexpected shades of emerald green in the shade. On my yellow bedspread one day, he turned a bright lemon hue in less than a minute, as if invisible brush strokes had gradually repainted him in front of my eyes.

His presence disrupted somehow the household's activities,

because servants refused to enter the room where they thought he might be. Our cook, Gathimu, and the house servant, Bitu, always avoided the study in which Emanuele's new friend wandered about freely, presiding over his homework in the afternoons, perched on books, scanning the ceiling for flies and mosquitoes, and stalking with silent glee the lazy, sluggish house-insects.

Many legends are linked to the chameleon in Africa, possibly derived from its mysterious mimetism. Africans therefore traditionally dislike them and prefer not to go near them. In the local legends the chameleon plays the part which, in the Bible, is attributed to the snake who tempted Eve in the garden of Eden: a strange conspiracy with the woman like whom it is variable, fluid, constantly changing into a fantasy of rainbows.

This dubious aura of 'untouchability' is what saves chameleons. It allows them to move around with impunity their defenceless prehistoric bodies, whose only natural enemies are snakes and birds of prey, who neither read books nor listen to stories.

King Alfred was the first of many. There followed various Marshall's chameleons, small and dark. A couple of rather fat ones, which were in fact named Fatty I and Fatty II: they did not have the protuberances on their noses and looked, more than anything, like hypocritical frogs. After them we had a Robert the Bruce, a Victor, a Kiwi, a 'Pembe Nussu' (or 'half horn' – it was mutilated) and King Alfred II. There were many more whose names I have forgotten. My son cared for them lovingly, letting them walk in the bamboos and on the gardenia outside my door, which attracted myriad insects with its rich perfume.

Then there was the time when three chameleons came with us on an expedition to Lake Turkana – still called Lake Rudolph in those days – because Emanuele refused to leave them behind. It was the hot and dry time before the April rains, during the Easter holidays. The journey from Nairobi took two days on the dusty tracks, and it was slowed further by our stops to look after the chameleons. Each time, their box was opened so that they could breathe fresh air, and be sprayed with cool water. They even managed to catch a couple of flies.

Yet the heat of the glove compartment in the Land Rover was extreme, overwhelming.

When, after hours of bumps, in the late afternoon of the second day, the breathtaking expanse of the lake, with its islands and shores of black lava and yellow grasses, appeared below the last bend like a primitive vision, Fatty I was dead.

In the open box littered with dried flies, his elongated body looked weirdly colourless, like the negative image of what it had once been. It had the temporary and precarious quality of frailness which belongs to small archaeological finds, discovered in the secret recesses of broken, desecrated sepulchres, which may disintegrate with the fresh air of day. It would not have surprised me if what was left of Fatty I had suddenly dissolved into pale dust.

This drama shadowed our jolliness. When, miles afterwards, we reached the oasis of Loyangalani, we left the car to have a drink and find relief in the shady breeze of the palm grove. Emanuele did not join us. He went off alone amongst the bristling grasses beside the path leading to the hot thermal springs.

When he came back he did not have the box, and his eyes glinted below the blond fringe. He had let the survivors go free in an environment that could allow their survival. Somewhere, under a lava stone, rested what remained of Fatty I. We matched Emanuele's sorrow with our silence.

Not so long ago, looking through his old yellowed papers, which have become precious relics to me now, I found a large blue exercise book roughly bound by himself. On the cover, a childish hand had written in red: 'My Chameleons'.

I leafed through it. It was dated July 1975, and it was written in English. Its tidy pages recorded, in his minute neat handwriting, names and dates, families, species, favourite foods and other details of every one of his chameleons; one of the pages was folded, wrinkled up and partly torn in small pieces.

I unravelled it carefully.

The passage was almost unreadable: It ended like this:

. . . Chameleons are extraordinary animals and fascinating hunters. I began to love them in 1972, and still now, in 1975, I love them. My favourites were Fatty I and Fatty II. They were very fast in eating, but slow in all their other movements. Whenever I let them go or whenever they escaped, I always found them again. Fatty I died of heat at Lake Rudolph.

On the word 'died', the pen had faltered.

Fifty Guineas' Pike

The seamed hills became black shadows . . . sounds ceased,
forms vanished – and the reality of the universe alone
remained – a marvellous thing of darkness and glimmers.
JOSEPH CONRAD, *Tales of Unrest*, 'Karain: A Memory'

Next full moon, I would like to show you and Sveva Fifty Guineas' Pike,' said my friend Hugh Cole. 'The sun will set and the moon will rise, and we will watch from the most fantastic kopje. The view is terrific. Pack some sandwiches. I'll bring the fishing rods. Yeah.'

He grinned. I found his Antipodean accent quite funny for a Cole.

I liked Hugh Cole. He was a true friend of mine, since the old days in Laikipia. The Coles lived, then, on Narok Estate, a large and efficient ranch situated east of us at Ol Ari Nyiro and, by Kenyan standards, they were our neighbours. Often Hugh and our friend Jeremy Block appeared with the specific aim of going after a buffalo with Paolo. They would be out all afternoon, and in the evening we would sit talking until late round the fire.

Hugh was not much older than a boy then – perhaps nineteen or so – and he had dreams, like boys have, and grown-ups, too, sometimes.

He was tall and lanky, with straight, dark hair inherited from his Irish ancestors, pale skin with some freckles, and a strange, veiled, deep voice. His most peculiar characteristic was the dancing look of mischief that crept into his disturbing blue eyes, which never blinked and focused on his interlocutor with the disconcerting fixity of a bird. Yet he was much too polite and far too well bred to stare.

Hugh had the flair for words of the June-born. His stories had colour and force, a cutting poignancy which I found entertaining, and in our long talks lay the core of our friendship.

Like the sons of the Delameres, of the Longs and of the Powys families, and of a few others, Hugh had been brought up to farm one day the vast family estates on the Kenya Highlands. They were families who belonged to Kenyan history, and to that earlier generation of eccentric, adventurous or aristocratic Kenyan pioneers who had walked their way through Africa against all odds at the beginning of the century, defying disease and heat, wild animals and tsetse flies, unvisited country and unfriendly tribes. Driven by the invincible curiosity to discover the unknown, they followed the dream of adventure and the need to explore intrinsic to the British soul, and they found a new Eden in the Highlands and on the plains of the Great Rift Valley, where they established their dominion.

They went everywhere on horseback, carving tracks and roads over virgin unwelcoming land. They died of malaria, of mysterious tropical diseases, of septic wounds and festering sores, of native spears or predator's attack. But they cleared the bush and tilled the fields. They bred prize sheep and pedigree cattle, and they shot the lions or rustlers that tried to kill or steal their livestock. They tamed rivers and harnessed springs, irrigated barren land and planted wheat and maize in hundreds of thousands of acres. Born within the boundaries, imbued with tradition, of a sheltered Victorian upbringing, they were in fact a tough lot.

Despite his inheritance, Hugh Cole's father in time decided to sell Narok to the new Kenyan settlers, as many people in the Laikipia Highlands did.

One day he gave Hugh some money and a pat on the back and he said to him, more or less, as Hugh himself years later narrated:

'Good luck to you, my son. Go safely. You are a Cole. You'll make your fortune.'

Slightly bewildered, Hugh set off to Australia and New Zealand. So did the English conquer the world in days past. But those days were gone, and it was hard for Hugh to find a place for himself in the new continent.

During a visit to America to meet his friend Jeremy Block, another exile, studying at a university, Hugh was involved in a horrific accident. On the bend of a mountain road in Colorado, his powerful motorbike flew off the cliff and while Jeremy, who was on the back seat, remained untouched and shattered only his watch, Hugh fractured most of the bones, big and small, in his body, and almost died.

In Kenya we heard about it, and were desperately worried for him. His recovery took years, and he never quite walked as before. But one day the phone rang and it was my friend Tubby, Jeremy's father.

'Guess who's back!' he said cheerfully. 'Hugh Cole. He is staying with me. Come to dinner.'

Much had happened to me in those intervening years. Paolo had died and my son too. But Laikipia was there in all its immense beauty, and Sveva my new angel, my child of hope and new beginning.

Although he limped and seemed to have grown slightly deaf, he was the same old Hugh, with his chivalrous manners and his well-told stories with outrageous twists; but there was a weariness about him, a new sadness, and of course the amusing hint of a New Zealand accent. He had done a bit of everything down under. Now he was back to see if there was anything more left for him to do here. He came to live with his sister nearby, and I saw him often, having picked up the threads of our friendship. We chatted, we laughed, we spoke of the old days, of people I had lost and he had cared for. My wounds were still open.

The invitation to explore was tempting, and the promise of

adventure always holds for me an irresistible appeal. I was curious about Fifty Guineas' Pike. So, on the given day, Hugh came up to Laikipia, in the bouncy green pick-up he used to drive like a maniac, in the back of which he always carried a couple of heavy cement bags to steady it. Sveva – who was then four years old – and I jumped in with a basket of sandwiches, and off we went. At Centre we gave a lift to Mirimuk, the head of our security guards, who wanted to visit some of his Turkana relatives over at Narok Estate.

Hugh had not been back since it had been sold, and of course he still remembered all the short cuts and the best ways through the old *bomas*, every detail of the place where he had grown up. He managed to keep a poker face and to show no emotion while we passed through the land for which he felt, I knew, such an attachment and which had been the background to so many of the stories he had told me; and I admired him for it.

We flew over pot-holes and rocks, biting the dust without mercy, as Hugh had always done. There was about him a new recklessness which was difficult to pin-point. I had no idea where we were going, and it seemed to me at times that even Hugh was no longer sure of his destination.

The landscapes we crossed were breathtaking. Undulating, green hills and open *mbogani* covered in low filigreed *acacia mellifera*, *sanseveria* and euphorbia. The country was much drier and more desert-like than Ol Ari Nyiro. It took longer than I had expected, hours of rough tracks and bounces, but when we arrived, the place was magical.

One of many kopjes which punctuated the landscape, it probably owed its unusual name to an obscure bet, the nature of which is lost to memory. Fifty Guineas' Pike was sensational and well worth the long journey.

At its foot there was a large pond with a waterfall rolling into it, surrounded by palms and immense, sheer, basalt boulders hung with wildflowers and papyrus. There were tracks of baboons everywhere and leopard spoor. Fish darted about in the pond, large silver barbel, looking like the classic images of fish that a child would draw. Water birds. Dragonflies.

We climbed up the kopje partly pushing and partly carrying Sveva, to the first platform which could be seen from below. There we found ourselves on a flat surface of smooth rock, dotted with deep cylindrical pot-holes, a geological curiosity possibly formed over thousands of years by the erosion of disappeared currents and swirling stones. From here the Ndoto mountains and the North Frontier on one side, and the vast expanse of the entire Laikipia plateau, up to Mount Kenya on the other, stretched as far as the eye could see.

Peering into one of the pot-holes, most of which held a brackish puddle at the bottom, Sveva discovered a small green grass snake, swimming weakly. It must have fallen in when searching for water, and was unable to climb out. We decided to sacrifice one of the fishing rods that Hugh had cut from a long thin branch, and which was rough enough for the small snake to wriggle up. We lowered it at an angle, and, after circulating it a couple of times, the snake started slowly winding its way up – to the sun and life and freedom.

'To Emanuele.' Sveva's small voice gave words to my thoughts.

'For Emanuele,' Hugh and I repeated.

The memory of my boy's lost laughter echoed again amongst the tall grey boulders. He had loved green grass snakes.

It was soon apparent that Sveva, with her short round legs, would be unable to climb the section of large rocks that led up to the flat stone platform, which was Hugh's goal, and offered hardly a grip even to us.

The sun was completing its arc in the sky. Nocturnal noises were beginning to creep in among the sounds of daylight. Hugh therefore decided to climb down and drive round to the back of the Pike, from where the ascent was easier. We drove back along our tracks, and at some stage left the beaten path, cutting across the bush for quite a while, heading in the Pike's direction. We parked the car in a small clearing next to a grove of acacia. Hugh took a water bottle and, leaving the parking lights on, we set off on foot.

The sun was fast approaching a chain of hills. Soon it would be dark and we had to hurry. Trusting Hugh's knowledge of the place, I did not think to memorize any landmark. The dry yellow grass was tall, the terrain sandy, fairly even, with scattered bushes. Not easy to find tracks here.

We followed narrow game paths, trying to keep up with him in the tall thorny vegetation which did not allow any view, and along corridors of spiky shrubs, winding their way slightly uphill, silhouetted black against the sky. Finally, we reached the back of the Pike and climbed it.

The view was spectacular. Magnificent horizons of craters, kopjes and hills, fading in the pale blues and pinks of the sunset, ran out to the foot of Mount Kenya, its peak covered in clouds, from which the full moon was about to rise. It was announced by the silver lining of the clouds and a pearly luminescence at the rim of the horizon.

Behind us the sun was setting beyond the mountains. Overhead, however, clouds were gathering fast to obliterate the sky and hide the rising moon.

The wind dropped. The clouds were here to stay. We groaned in disappointment. Still hoping the sky would clear, we chatted, drank some water, sang a song. Suddenly it was pitch dark.

It was soon apparent that the moon would not become visible for hours tonight. Baboons began to bark their goodnights from the sleeping cliffs; alarmed goodnights, for rather close we heard the unmistakable, rhythmic, rasping voice of leopard. If I had been alone, I would not have hesitated and would have slept there, safe on the flat warm rock. But Sveva was tired and would be hungry, and it could easily rain.

We descended from the platform by the way we had come,

tentatively now, Sveva gripping my shoulders, and we landed in the thick bush. We then tried to find our way back to the car, and it was a mistake. I had a pocket torch, but its tiny light, absorbing and distorting every shape, made the encircling night darker, vaster and confusing. Every bush of *mellifera* looked like every other, hung with powdery yellow flowers unendingly identical. Every next turn of the sandy game path seemed like the last one. The hills were now invisible, and with no reference we walked fatally in circles.

Finally Hugh cleared his throat and turned to me. I heard his deep disembodied voice spell out from the shadows what so far we had not dared to admit.

'My friend, I am afraid I have got you lost. I am sorry.'

At the word 'lost' Sveva wailed.

I had never been lost before. I was surprised how unsettling and undignified the very thought was. Ideas came racing through my mind and failed to find solutions. I became impatient and angry in my dismay. The anger was mostly annoyance with myself, at my stupidity. I should have found out more, looked around better and not allowed such a ridiculous and unnecessary situation to happen. Getting lost indeed. I took hold of myself.

'You brought us here, and you are going to get us out of here,' I coolly told a crestfallen Hugh, trying to convey in my voice a calm and flippancy I was far from feeling.

Sveva's hands clutched my shoulder:

'I want Wanjiru,' she declared with a hint of defiance in her trembling voice. 'And the *askari* and my room. And I want Morby.' Morby was her beloved soft pink mouse. She must have felt immensely remote from her safe known world, hanging from my back, lost in the African night. I tried to reassure her:

'We shall find the right track at any moment. It is great fun to be here. No little girl we know was ever as lucky as you are. Imagine, a real adventure to tell your friends. Now you must help

me to guess where to go. In the meantime, we shall find a cosy
place to wait.'

Many times, while looking for lost cattle in thick bush and
rocky terrain, with no visibility even in daylight, I had seen Luka,
our tracker in the old days, smell the wind as he followed a track
he could not see, turning his head here and there and going off
in an unlikely direction which was infallibly the right one. I had
been curious to know how he could do this, and often asked him
to tell me how he managed. He had looked at me in puzzlement,
for following an instinct is something impossible to explain.
Invariably he had said: '*Lazima jaribu kufikiria kama nyama,
memshaab. Ngombe hapa wataenda kulia kufuata arufu ya maji –
hama kutoroka arufu ya simba.*' 'Try to think like an animal,
memsaab. Here a steer would now go right, towards the smell of
water, or away from the smell of lion. Cannot you see?'

So I tried to think like an animal, which meant following my
instinct, without resorting to my reason at all. It told me to turn
and go in a certain direction. Shortsighted and in the dark, God
knows how I could. I prayed we would not meet a buffalo coming
to water.

We advanced tentatively, often to find thorns tugging at our
clothes, pulling our hair and scratching our bare legs. I tried hard
to see the bright side of things. At least it was not pouring with
rain; at least it was quite warm, and the air was balmy with heady
scents. It was a beautiful night, after all. I was in Africa where I
had always wanted to be and this was a mysterious place, undisturbed by people. This *was* the quintessential adventure. At every
turn of the narrow track I strained my ears to anticipate a grunt
too close, a rustle of leaves.

Amazingly, a white and wider track suddenly opened in front
of us, and with unspeakable relief I realized it was the road. But
at which point on it? And where was the car – right or left?

The only wise decision was for Sveva and me to stay right here,
with our back to a large shrub for protection, to light a fire to

discourage nosy beasts, and to wait. Hugh would go on looking for the car. If by sunrise he had not returned, I would set out on my own to find the car tracks. There was no sensible alternative. Staying where we were was safer than wandering about with a small child in the darkness near a watering hole. I was certain that all sorts of animals would come to drink there in the night; I remembered the spoor we had seen.

We gathered some sticks and larger branches by the light of my little torch which was beginning to glimmer only tremulously now, and Hugh lit a fire with a dexterity learned in the free days of his childhood. The orange flames leapt high, crackling, and the shadows receded to a wavering circle with borders of mystery from which countless eyes seemed to peer at us.

The fire lifted my spirit, and Sveva lay on my khaki sweater, quiet now, and listening, like I was.

Hugh stood a moment and took his neck scarf off.

'I shall tie this to a bush at the point where I leave the road.' He grinned once. 'Are you sure you will be all right?'

He bowed slightly and was gone.

Around our small fire, and its warm halo of light, like a dot in the universe, the night was black and vast and alive with unknown, magnified noises. Red and brown ticks sidled away fast from the flames, running diagonally like minute crabs. I kept telling Sveva stories, without a pause, to cover the yelling laughter of the hyena, and the threatening voices of the invisible and of the unknown. I sat with my child out in the African night, which for once did not belong to us, feeding the fire with thorny branches of *mellifera*, and trying not to think, or rather planning what to do, if the lion which we heard roaring from the hills decided to come to drink at Fifty Guineas' Pike. My ears strained to catch all whispers, to interpret secret shuffles, a sudden roar that froze my throat for a moment, trotting of heavy hooves along dry paths.

My Italian past was far away that night. Yet there was a primal beauty and privilege in being here as we were: where was the rest of the world?

With the infallible knowledge of a child born and brought up in the bush, Sveva broke the silence now and again.

'Mummy, hyena.' Something had yelled.

Or, on a close trumpeting: 'Elephant!'

Something coughed: 'A leopard!'

A burst of loud, untidy barks: '. . . and he is eating a baboon, Mummy.'

She was probably right. African noises can tell their story far more eloquently than human words.

Looking at the fire and nursing it, time ticked by with the ancient sound of crackling embers until, my watch forgotten, my head lost in thought, a peace descended, and the awareness of being safe. A total calm and the feeling of being in the right place.

Gradually, and then abruptly, the great clouds parted like waves, and in a crystal-clear sky a cool full moon sailed silent and aloof above, ignoring us. It was hard to believe that this was the same moon people were watching at this same time from a London window or a Venetian gondola. The unknown hills were clear-cut against a translucent sky, and a deafening concert of crickets and frogs erupted to greet the moon. The nearby sounds were friendly, and soon became a concert of which we, too, were a part. There was a silvery magic about the place and us, alone and vulnerable, yet accepted by the creatures of the night. I let the fire quieten down, twinkling, subdued and no longer needed.

I took Sveva in my arms, her warmth and child's fragrance comforting and safe. I knew well that I would never again capture the essence of Africa as I had tonight. Time went by. I was happy. And it was with regret, not relief, that much later I heard the muffled hum of an approaching engine and car lights put an end to the spell.

We laughed with abandon, like reckless children. I covered the glowing cinders with sand and got into the car, but as the sun began to rise and the familiar shapes of the Laikipia hills came into focus, we grew silent. Sveva was sleeping, and the adventure receded like all dreams at dawn, to become a memory.

Upon the Wings of the Wind

In memory of Tim Ward-Booth

Yea, he did fly upon the wings of the wind.
Psalms XVIII, 10

L ife in Kenya, with its extraordinary beauty and variety of
opportunities, its unbounded space and spectacular landscapes teeming with wild animals, its lakes and deserts, mountain
ranges and countless beaches, its savannah, forests and windswept Highlands, attracts people of an unusual quality, who
regard risk and challenge as an intrinsic part of the *safari* of existence. They fly with the moonlight and land in the dark; they
hunt alone for lion and buffalo in thick bush, or for crocodile,
wading waist-deep into rivers and lakes; alone they climb deceiv-
ing mountains; they explore on foot waterless deserts and forbidding country where bandits are known to attack travellers; they
dive in shark-infested waters, or sail with light craft in turbulent
seas; they defy malaria, yellow fever and tropical disease; they
approach dangerous animals to study or film them. They gaily
court danger, and, although a number of them perish, quite a few
manage to survive.

But for too many, as for some who live more conventional
lives, the end of the adventure is simply a road and a lorry which
does not stop.

Kenya's tarmac roads are notoriously unsafe, populated as they
are by unworthy vehicles and irresponsible drivers, who race at

full speed careless of rules, leaving chaos and tragedy in their wake. Among them the worst by far is the Mombasa road, where hundreds die every year in terrible accidents, the majority of which could have been avoided. So died Paolo, and dozens of people I have known. One of them was Tim.

When I think of Tim, now that his time on earth has gone, and his body rests on a hilltop overlooking the desert of Kenya's Northern Frontier, as close to the sky as when he lived, what I remember most is the beat of his last helicopter, and that of my first. It was an afternoon years ago, soon after we had met, when he flew me down through the shadows of the Mukutan gorge.

The deafening noise of a helicopter's propeller always reminds me of a struggling gigantic insect flapping its wings in a last frantic attempt to fly. Over that noise his voice sounded clear and deep.

'Are you ready?'

He turned towards me, grinning. I noticed the curls on his straight head, the handsome and masculine features, the Roman nose, reminiscent of Paolo's. Like all men who fly, his eyes had a different quality – penetrating and yet at peace, shining with the innocence and cleanness of the space over the prairies of clouds, infinitely remote from the polluted world of the crawling creatures below.

There was this aura about Tim that I had only sensed before in men I remembered long after the echo of their steps had faded down the corridors of time. Men who did not live for long, whom I could not imagine growing old. It was a presence both warm and aloof, strong yet gentle, which commanded instant respect. He did not talk much. He moved straight, with no effort of his long lean legs, his skin tanned golden by the Equator's sun.

I looked out of the convex glass window, at the sheer cliffs of rock hung with aloes and thorny euphorbias, and down at the dense carpet of palm tops and fig trees which covers the bottom of the Mukutan gorge. For years I had wanted to explore the depths of the canyon, which are impenetrable to humans, fit only for eagles, vultures and daring helicopters of silver.

The occasion was during the period, soon after Emanuele's death, when Robin had come into my life. He was working on a film about a jungle adventure which involved a scene with a helicopter, to be shot at Nyahururu, formerly Thomson's Falls, about forty miles from Ol Ari Nyiro. Tim was the helicopter pilot. I went to visit the set, and in the evening I invited Tim to spend the night in Laikipia with Robin and me, and to fly us down the Mukutan gorge. He was quick to agree, and we went.

I was ready and nodded. My heart beat hard in my head, and there was a rush of heat to my face, as we began to plunge into a vertical dive, encapsulated in that precarious metallic machine which sounded so much like a crazed dragonfly. I looked at the back of Robin's head. He was in the front seat. His neck hardened in tension, a rivulet of sweat trickling down into his blue shirt.

The gradients of rugged rocks were so precipitous, so mercilessly steep that only Tim's skill, refined in the time when he had served in the Falklands war, could carry us through the narrow passages. But I had suddenly no fear at all. He was totally at one with his machine, poised and completely in charge, as balanced and light and precise as the bird perched on a slight swaying treetop I had once watched with wonder in the Seychelles.

Tim concentrated on the command panel, his ancient profile still and timeless like the portraits of warriors engraved on Roman coins. His eyes narrowed, and we were dropping along the sheer walls of pink and grey stone towards the tufts of the treetops 3,000 feet below. I felt the same mixture of elation and worry, of being physically winded and mentally exhilarated that I had as a child, on my first wild swoop down the highest, steepest slide at the Luna Park.

The green leaves were suddenly too close, almost touching the keel. For a few moments we were tentatively brushing the treetops, like a bird searching for a safe branch to perch on. Then we glided horizontally along the narrow bottom of the valley, past waterfalls, tangles of liana and dracaena, monkeys and eagles and quiet unvisited ponds. Soon we were surging again up high, to the top of the canyon, where boulders of granite had for millennia guarded the silence and mysteries known only to the African creatures. There we emerged into the different windswept world of the Highlands plateau, the vast limitless expanse of the ridges of Jaila ya Nugu, Nagirir, Kurmakini, Mlima ya Kissu and the familiar favourite promontory of Mugongo ya Ngurue.

The sun was swallowed by the valley, and the shadows spread fast on the hills. We had veered obliquely towards my house at Kuti, low over the bushes, scattering some outraged elephants, while buffalo looked up, rooted on their stocky legs, more puzzled than aggressive for once.

We landed on my lawn. The staff and all the dogs had gathered together in a bewildered group to watch in awe, from a distance, the flying object from another star descending in the dusk. They clapped their hands and jumped up and down and laughed on discovering that it was Robin and me who disembarked, running, our hair windblown, with Tim following. And he was laughing too.

Years later came the afternoon at Lewa Downs. This time I stood at Aidan's side. Tim had been his cousin. The scene was unreal, the beauty and the pathos of Africa were at their most dramatic.

In the haze of noon, one after the other, cars of all descriptions drove up the hill. They parked in neat lines, just below the four army helicopters which rested incongruously on the grassy slope. People got out in silence, and climbed up to the summit, where half a dozen armed African rangers in khaki-green uniforms stood to attention, a depth of sadness in their still features. The women's skirts and hair moved in the light wind. In the pure sky, unblemished by clouds, a solitary vulture flew, close to the sun.

A few giraffe ambled quietly to the shade of a yellow fever tree down in the valley, and to the north the views of pale lilac hills stretched ahead unendingly to the distant horizon.

From a four-wheel-drive car emerged a girl, dressed in black. She had been driving back to Nairobi on the Mombasa road and there she had found a body thrown into the middle of the tarmac – the killer lorry gone. It had just happened. It was Tim. Courageously she had put him in the back of her car and had driven him straight to the hospital in Nairobi. Only in Africa can such things still happen.

She moved towards the lady, dressed in white, who was his mother. I introduced them and watched in respect while they embraced. The first woman to see him alive, and the last. Encounters at the limit of human experience.

People met like this on that hilltop at Lewa Downs on the afternoon of the burial. People who had come from far for this last goodbye; people who knew each other well; people who had never met; some who would later become friends; others who would never meet again. All of them, tomorrow, would go back once more to their diverse, separate lives.

Then the silence was broken by a strange pulsating sound coming from the sky. I looked up. The vulture had gone, and here, as unexpected to me as the unicorn of legends, was another helicopter.

It landed on the peak. Music started, like a lament. The doors opened. All the men took off their hats, and his friends approached to take his coffin. They carried it in silence to the very edge of the hill, where they placed it among garlands of heather from the moorlands of Mount Kenya.

Suddenly, with a tremendous noise of flapping wings, almost from nowhere, two more helicopters converged simultaneously on the hilltop, and for endless moments they stayed still, suspended over the valley, facing the coffin and level with it in a glorious salute, while a man in uniform, tears rolling down his cheeks, played the Last Post.

Part IV

A WORLD BEYOND

The Cobra Who Came
from the Dark

Todo dejò de ser, menos tus ojos.*

PABLO NERUDA,
Cien Sonetos de Amor, Noche XC

T he feeling lingers of a drifting young presence below the
shadows of the yellow fever trees in the night garden. A voice
comes sometimes with the voice of the wind, joining the starlings'
song in the full moon and the crystal calls of the tree-frogs from
the fishpond. A faint voice, like the echo of my memories, and
the unreachable essence of dreams.

It is then, when I am alone in the heart of my home, protected
by my dogs sleeping content around me on the carpet, the fire's
flame subsiding into breathing, flickering embers, that he comes
back. In the quiet room, behind my desk, there is suddenly a
presence. I do not turn my head. I keep watching the enlargement
of the photograph hung on the wall, the one taken by Oria after
that last picnic at the springs.

Forever he goes away up my wall, leaning forward, in the red
dust of his motorbike, driving fast towards the hills, without
turning his head to look back.

I search, sometimes, when I walk alone in the sunsets, for his
lost voice, for phrases already spoken and forever gone. I recall
how he walked, how he shook his head to clear away the hair

---

* 'I could see nothing more, apart from your eyes.'

from his forehead or a thought from his mind. But the image comes and goes away too quickly, dissolving before I can quite manage to grasp it and savour it, before I can fix it in the hourglass of my present. I cannot smell any longer the warm smell of his young boy's skin toasted by the sun, and it is only the scent of buffalo and elephant, of jasmine and sage, which mixes with the breeze from the east, blending with the sound of birds and the rustle of alarmed gazelles and jumping hares.

What remain forever of him are only his eyes.

From the drifting fog of my vision only his head seemed to emerge; his mouth curved slowly in a spreading luminous smile, and the firm unwavering eyes kept looking inside me unblinkingly, yet without staring. The background seemed to take on an intense, unearthly nuance of vivid blue, vibrant with a cosmic quality which I perceived but could not explain. A smooth shape, with glistening marbled skin and beady, lustrous eyes, coiled easily just below his chin.

I stirred.

'I can see him. He seems serene. He looks peaceful. And he is serious. Strange. He . . . has a snake around his neck.'

The lady in the red sari smiled. Her short white hair was cut like a man's, but a lock touched the red dot of paint forming a fine upward pointing arrow in the middle of her forehead. The eyes glowed bright like burning coals and they transmitted care and warmth, mixed with the unusual gift of a totally accepted and mastered insight. She put her hand on mine, and instant heat radiated from the dry palm. Her brown fingers, adorned with silver and gold rings, rested a moment on mine, and I felt a surge of great peace, an encompassing calm overtaking me. I closed my eyes and, even before she spoke, I knew that what she was going to say would be the answer to so many of my questions, and that it would be right. The weird, uncanny tie between Emanuele and whatever was Indian.

'Breathe deeply. Yes, Kuki. But to finish his karma just do what I tell you, when you return to Kenya. It does not matter if you believe it or not; one does so many things without knowing why. It will not hurt to do it. With Sveva, the first night you are back.'

Outside the tall window, the great mountain peaked with snow, whose shape seemed to me inexplicably familiar, hovered benign and extremely powerful. I had heard that some people found it threatening, and oppressive, and fled from it in fear. To me it was a protecting strength, fully deserving its old Red Indian name 'The Mother'.

The shadows of a setting, late-summer American sun advanced slowly, and the granite boulders of the summit were suddenly tinged a deep blood red. For the first time I understood the reason for their name, Sangre de Cristo Mountains. And their blood was my own, and everyone's, the sadness of the universe, and Emanuele's blood.

The milk filled the cup. Sveva withdrew the jug she was holding and looked up at me.

'I think it is enough. Let us go.'

We went out into the darkening gardens, followed by our patient dogs. It was another sunset, short and of drifting blues and reds and greys.

Our guests had retired for an evening shower and to change for dinner.

Sveva and I were alone, and this was the right time.

At the end of the garden the fire the *askari* had just lit was gaining strength, and its light, blending with the fading glow of the setting sun, was not yet in contrast with the darkening night.

The dogs ran ahead, chasing each other, and looking back curiously at the wooden bowl Sveva was holding. Together we managed to fix it, without spilling a drop, in the fork of Emanuele's tree, so that the dogs could not reach the milk.

We stood side by side as we had been told, murmuring

together the words that the Indian priestess had taught us, the haunting Hindu sing-song of the Vedic mantra. Strange, soothing words we could not understand.

Their sound, their purpose, suddenly struck me as being of the same essence as any Christian, Muslim or Jewish prayer, or any pagan invocation, the instinct of humans to try and reach out to touch the inexplicable infinite that we call God. Their unknown meaning, by rendering them arcane, magic, and in a way more plausible, matched the incomprehensible mystery of death. For different reasons, in different languages, but with similar rituals, millions of people were praying, at that very time, with the same hope to the same Unknown to whom they gave different names. Standing on the Highlands of Kenya in the gathering darkness, the Latin murmurs in the churches of my childhood were closer to me than they had been in a lifetime.

Om's echo drifted away, floating between the wings of the wandering nocturnal plovers, whose voices sent a rain of arrows to the sky. The fire burned fiercely, fed by the wind. I touched both trunks in a lingering caress; the soft yellow fluff of the fever trees planted on Emanuele's and Paolo's graves trapped the warmth of the sun like a human body.

Sveva and I walked back, holding hands. In the fishpond a half-moon was reflected. But amongst the papyrus, where the little silvery tree-frogs hide, inflating their small throats like pearly bubbles in a trilling song of bells, there was no movement. We searched the still oily water with our torch, but we could see no goldfish gliding away lazily below the waterlilies, under the carpet of salvinia. The fish pond seemed strangely silent that night.

We went back to the house, where another fire had been lit on the fireplace.

Early next morning, devoured by curiosity, still with our nightgowns under our caftans, Sveva and I went to look. Nothing seemed to have changed in the night. Even from a distance, we could see the wooden bowl lodged in the same place, between the arms of the smaller fever tree. I took Sveva on my hip, so that she, too, could look at the same time. Small sugar ants busily climbed through the fluffy bark. Up in the intertwined branches filtering the morning sun, a starling perched, and flew off. We looked, holding our breath.

The bowl was empty.

We walked back in total silence, not daring to give words to our thoughts which had no answers. Sveva threw a piece of her morning toast into the fish pond. A normal action. The bread disturbed the floating weeds, ripple after ripple unbalanced the blue water hyacinths, but no fish or frogs dashed up from the bottom to grab the food. The bread remained uneaten. All life seemed to have abandoned the fish pond. It was puzzling. I went to look for one of the gardeners to ask if any ibis or stork had come to eat the fish.

The one *shamba* boy I found had not seen any.

It was dark when we came back that night, after a long game drive with the car full of guests and friends. We were commenting on the troupe of baboons always perched on the tallest trees at Marati Ine, just above a rotten trunk where Emanuele had found his first cobra. The gardeners were waiting at the outside gate. I stopped the car to ask what was the matter.

'*Bunduki.*' 'A gun,' said Francis. 'We wanted to ask permission to call a watchman with a gun to shoot the snake.'

A feeling of hair rising at the nape of my neck. A feeling of having looked down into the unknown, to discover a familiar, yet indecipherable link.

I switched the engine off while everyone was listening.

'Which snake?'

'The one who came last night. We saw him at the graves, but he disappeared into the fish pond.'

That explained the eerie silence.

'Which type of snake?'

I found it difficult to find my voice. Sheelah had told us which. The snake sacred to Shiva, the most deadly, the most holy of all.

'*Kiko,*' said Francis. A cobra. Then his voice lowered to an awestruck murmur:

'But we have never seen the like of him around here. He is the large variety. The one which stands straight upright.'

The giant forest cobra. The king of cobras. The most holy of all.

'Mamma!' Sveva was shouting in the waiting silence. 'Sheela had told you. We must let her know. It worked. The mantra worked. Emanuele is all right now.'

'No gun,' was all I said to the gardeners. 'Leave that snake alone. As he has come, so he will go, and he will follow his ways and the ones of his spirit god.'

Next morning we went to the fish pond, Sveva and I. Weaver birds were flying in and out, busily ripping the papyrus into shreds for their nests. Midges flew in a small cloud and dragon-flies darted jerkily from a creamy waterlily to a blue one. When Sveva threw in her bread, dozens of fishes of all sizes sprang to life from the muddy bottom, fighting for the bobbing morsels.

The cobra had gone. The soul could rest.

The Pendulum

Home is the sailor, home from the sea.
                    ROBERT LOUIS STEVENSON, *Underwoods*,
                                           I, XX, 'Requiem'

He was a pleasant man with old-fashioned manners, tall and still good-looking in his middle age. His greying hair, impeccably parted on one side, set off his tanned skin, regular features and dark, lively eyes.

Like Dickie Mason, the father of Emanuele's friend Charlie, he had been in the Royal Navy, and of the dashing young officer he still kept the upright bearing and chivalrous ways. He loved sailing and when we went to Kilifi he often spoke of long-gone adventures of the sea.

He excelled at telling stories and had, in fact, a fascinating and seemingly unending repertoire of legends and tales of the coast. Most had to do with the magic beliefs, fetish baobabs, full-moon ceremonies, and the happy or unhappy ghosts for which Kilifi, like Takaungu, Vipingo, Mtwapa, Shimoni and the other Kenyan creeks, was renowned.

In the not-too-distant past, Arabs still sailed to East Africa to fill their dhows with ivory and game skins, spices, coconuts and slaves. Before loading them at night with the high tide, they would chain them to the walls of the ocean caves along the fiords that abound along the Kenyan shores, safe mooring for sailing boats, out of the reach of the ocean's swell.

The spirits of the slaves linger around the many Muslim ceme-
teries for which the coast is well known, and amidst the ancient,
mouldy ruins of abandoned towns to be discovered in the forest,
among the roots of giant bread-trees and baobabs, overgrown
with luscious vegetation, liana and wild orchids. The local people,
the Giriama, still kept alive their witchcraft and secret rituals.

I liked all these legends, as they belonged to the atmosphere of
mystery and exoticism of the coast, the strange ripe scents of
wild jasmine and frangipani, cloves and cinnamon, vanilla and
incense, sandalwood and musk; the unusual rich taste of caramel-
ized pineapple, star-fruit, *madafu* and sharp lime; the prevalence
of a thousand dangerous snakes and large iguanas, prehistoric
monsters with eyes like glass beads, and flickering tongues; the
innumerable coloured birds and monkeys, fruitbats and bushbabies; the solemn baobabs, like columns of vanished temples, and
the palms and casuarinas forever shivering with the salty breath
of the monsoon.

I had known him for as long as we had been going to Kilifi.
Yet although in Paolo's and Emanuele's days we saw much of
him whenever we went to the coast, he was really a social
acquaintance who had always retained a certain aloofness, and I
could not say that, although I liked him, I had known him well.

I was therefore slightly surprised when, in the obscure days
after the disappearance of Emanuele, a very sensitive letter of
condolence was followed by the message that he would come and
visit me in Laikipia for a week.

Emanuele had been dead only three months or so, and the
pain of his absence still lingered, shrouding me in waves of solitude and longing.

I had been suffering from a sinus problem, created by the dust,
and when he arrived I was still nursing a searing headache. He
looked at me and, with half a smile, he dug into his pocket and
produced a strange object which he swung in front of my eyes. It

was a cone of brass, polished to a bright shine, that had been tied
to a length of fishing line. I must have looked puzzled.

'Do you know what this is?' he asked. 'A pendulum. It can be
used for a variety of purposes. Particularly to heal, but also to
find something that you have lost; to get an answer to a problem;
and to discover water. I can try to help you. Then perhaps I shall
teach you to dowse with it.'

I was intrigued.

He held his pendulum above my head with firm hands, and
sure enough, in a few instants it began to swing, slowly at first,
then faster and faster, almost disappearing in a rapid swirl. Just
watching it I felt dizzy: then the pain in my forehead seemed to
be lifting. In a few minutes I felt much better. I was genuinely
amazed.

He suggested that we should try it a few more times over the
next day or two, and after that I was totally cured.

'I wish I could do the same. It must be wonderful to be able to
help people in such a natural way.'

'You probably could,' he told me seriously. 'Most of us have
this capacity, if we would only care to develop it. In the meantime
let's go and look for water. We shall see.'

I was at that time trying to occupy myself with the works of
the ranch and farm, and needed to locate some places at which
to dig bore-holes. His unexpected help was most welcome. As a
farmer, he was very knowledgeable about crops and cattle, and
was extremely interested in our activities at Ol Ari Nyiro. There
was a calm and a depth about him, and it was fascinating to go
round the ranch with him, identifying the best locations for a
bore-hole.

We would leave the car and set off on foot until we reached a
spot that seemed suitable. There we held the pendulum in front
of us without moving, waiting for it to rotate or swing on its
own. Often nothing happened, but sometimes the reaction was
so strong that I was taken by surprise. In this case we double-

tested the spot with a rod cut out from a shrub. The first time that it began to vibrate, almost jumping from my hands, and then to point forcefully towards the ground, I was caught completely unawares and was so elated that I almost dropped it. It felt like stepping into a different world, where the unknown forces of the earth that had guided past generations revealed themselves, in their uncanny natural strength and at the same time in all their simplicity.

I was an enthusiastic pupil. Time flew, and I was sorry when the day came for my friend to leave. Just before getting into his car, he took the pendulum from his pocket.

'I would like you to keep this,' he said, looking at me seriously. 'But you must promise that you will use it. You have this power, and it is your duty to practise it.'

I protested. I knew how much that pendulum meant to him and I could see that a sort of personal and intimate tie existed between the two of them, rather like that of a wizard and his magic wand, or a witch and her black cat. But he put it firmly into my hands, and, before I knew it, he was gone.

I used the pendulum now and again as time went by, mostly on Sveva when she was sick, and once or twice on my mother and some of my friends. Many felt a certain relief, and I could never work out how large a part suggestion played in this exercise. Most people who knew him remarked how extraordinary it was that he had given me the pendulum, as it was well known how much he valued it. I felt proud to have been chosen.

I did not see much of him over the next few years. I never went down to Kilifi again, as it held too many memories of happy days, and I knew he came to Nairobi only occasionally.

One morning, I sat at my hairdresser in Muthaiga, and was reading a book I had just bought, because the title had triggered my curiosity: Mysteries, by Colin Wilson. The first chapter was called 'Ghosts, Ghouls and Pendulums'. Inevitably, I thought of him, and wondered how he was. We had heard that he had remarried; it had been years since I had seen him.

It was a sudden sensation of being watched, and I lifted my eyes. There, amongst the ladies in curlers and the bottles of shampoos, in the mirror in front of me, I saw him. At the same time he turned his head, our eyes met and he smiled in recognition. I was so astonished by what I could not call a coincidence that for a moment I could not talk. Instead, I help up my open book towards him, so that he could read what I was reading. His smile simply widened but without a fraction of surprise, as if what happened had been totally normal.

'With you it was to be expected,' he said calmly. 'Are you well? How is the pendulum?' I noticed he looked tired, almost breathless. It was a hot day.

Only a few weeks later at a dinner party in the Belgian Embassy, someone mentioned casually: 'How tragic, what happened in Kilifi.'

And this is how I heard. He had left his house one morning, with his shotgun, and went off to Takaungu for a walk. He never came back. They had found his headless body in the afternoon, the gun close by, on the stretch of beach along the creek not far from Denys Finch-Hatton's old house.

Nobody could explain what really had happened.

He had gone, this loyal friend of past days, with his mysteries and his stories, a secret pain never to be known. Another Kenyan drama, another ghost to join the ones of Kilifi under the baobab trees.

When I went home I looked for the pendulum, but I could not find it where I had left it. I searched everywhere, for days, combing the house and all possible hiding places. The pendulum seemed to have vanished.

It was only recently that I met his widow for the first time: a good-looking lady who could still not believe what had happened. She wanted to write a book about her life and hoped I could help her with publishers' addresses. She knew I had known her husband, but no more.

At one stage in our conversation, I could not resist asking her about his passion for the pendulum. I was about to tell her that he had given it to me, that I had been very honoured, and that I was now shocked because I could not find it, when suddenly she said:

'About his pendulum. I threw it into the Indian Ocean. It was the only right thing to do.'

The beach, again, was left only to seagulls.

The Road to Rubu

We shall not cease from exploration
And the end of all our exploring
Will be to arrive where we started
And know the place for the first time.

T. S. ELIOT, *Little Gidding*

O ne summer, not so long ago, we decided to go north of Kiwayu to explore, Aidan and Sveva and I. We bought a second-hand Land Rover old enough not to represent a temptation to *shiftah*,* but in good enough condition to carry us safely with the luggage and camping kit that we had purchased. Aidan also acquired a boat, a sturdy white plexiglass dinghy with an awning to shelter us from the sun, a good engine, and a spare engine, which had been Paolo's, kept safe in a store in my house at Kuti, where it had waited for years, for the time when again it may feel the same waters.

The plan was to drive north of Kiwayu, cutting a track along the overgrown trails to Rubu, on the north Kenya coast. This was an area which, for over a generation, no one had gone through because of the constant threat from bandits roaming the lands near the Somali border, around the frontier town of Kiunga.

I had journeyed there first in the summer of 1973, when Paolo and I, young, in love and eager to explore remote corners of the country we had just chosen as our home, had travelled north with our car and a rubber dinghy, and spent two magical weeks in a tent pitched on the beach at Mkokoni.

*Bandits from Somalia or Ethiopia.

To reach it we had driven through barren and untamed country, encountering wild animals, visiting small Muslim villages and making friends with the people. With our boat we had glided happily round the canals and climbed up steep tangled slopes, spotting wildlife on the sandy dunes covered in sea vines, and marvelling at the rare privilege of discovering that so much virgin, unspoiled land still existed, it seemed to us then, for us alone.

So it was, to some extent, to retrace our steps, that I returned twenty-three years later, and found that very little had changed.

The elephants were no longer there, belly-deep in the tidal ponds, and the blue waterlilies had died out. But we saw a cheetah the first day, and lesser kudu and dik-dik at the side of every track.

We drove into Mkokoni first, looking for Mote, the fisherman Paolo and I had employed that long-lost summer, more than twenty years ago.

He had been a young man then, nimble of body, quick to spot fish in the shallows, agile and ready to steer his boat through the labyrinth of mangroves of which he knew every turning. In those two weeks we had shared adventures, and, with unfailing generosity, he had put himself and all his competence at the service of these two young *wasungu*, the first to come so far through dangerous unvisited country, and to stay on.

Over the years I had not forgotten him, and I had seen him once in the late seventies during one of Emanuele's half terms, when an exclusive safari village for discerning guests had been built along our beach, and I went to find some of my memories and him.

Emanuele had loved to collect cowries in those days, and Mote promptly offered his *maho** and two of the crew to sail out of the

*A smaller version of a *daho;* traditional wooden sailing boat used by fishermen in the North Coast.

Melango at high tide and wait for the tide to go out, exposing the bank of molluscs at Oyster Bay, where, with luck, a *Cipraea Vitellum* or perhaps *Caput Serpentis* could be found, to my son's delight.

I had not seen Mote since, and was not sure if he was still alive: after all, Paolo and Emanuele had died, wiped out by an indifferent destiny, and the life of a fisherman in the Indian Ocean is not a long one: his brother had been eaten by a shark while diving for lobster just out of the Melango.

But Mote was still there.

He came in from the beach where he had been fishing, a loin-cloth round his belly, a live lobster glistening in his hand, still dripping wet and with the light step I remembered, small wizened face, and the slight figure of an old child. He looked up at us searchingly, and his face split almost instantly in a delighted grin.

Years ran like sand through the hourglass of his memory, and I knew he had remembered me.

'*Allah Hakbar! Mama Kuki! Kwisha rudi! Leo mimi nafra. Nashukrani kubwa sana kwa Mungu.*' 'May God be praised. Mama Kuki is back. I rejoice today and give thanks to God.'

He intoned these words in a high-pitched voice, jubilant, and his cries soon attracted half the village. His brother Haji came out, bearded, older seemingly, fatter, greyer, resembling our Mote not at all: by another wife, I wondered?

Bajuni children and beautiful women assembled, colourful and elegant, greeting us. Soon we were sitting in the shade of Mote's house, on the Swahili beds of thatched doum-palm leaves, and were offered *tangawezi chai*,* with no milk. It was the full 'travellers' welcome', traditional in this strict Muslim community. Shelter, drink, food. Small sweet yellow corn cobs were produced for us, while we told of our safari and of what we wanted to do. A young man was recruited from the village, and our boat

*Ginger tea.

on its trailer was examined by everyone, found good enough and accepted.

It took us three days to reach Rubu.

The village had been abandoned over twenty years ago, when constant *shiftah* incursions from neighbouring Somalia made it impossible for the inhabitants to stay on.

The track had become overgrown, tangled with liana, sansevieria and thorny acacia shrubs which scratched deep ridges on the car's metal, impeding our progress. But Aidan was determined, and with Mote, Ali the boatman and our young recruit, plus the drivers we had brought with us, we managed to clear a large enough gap for the vehicle to proceed slowly. We camped at night in the middle of the track – hooking mosquito nets on bushes over our mattresses – so dense was the vegetation around us that there was no room for our tents.

The bleached bones of an old elephant marked a spot, below five tall baobabs, where the entrance of Rubu had been. There were the crumbled ruins of a mosque. Small heaps of porous, decaying coral rocks indicated where houses had once stood, and disintegrating wells, made of yellow sand and lime. Arab inscriptions from the Koran, engraved on a few oval tombstones planted around with blue lilies, discovered behind a dune, indicated where the Rubu community had once laid their dead to rest. To one side, still partially functioning, was a flat *jabia*, the ingenious traditional way to collect rainwater to drink.

Sand vines and sansevieria grew everywhere, and on the slim stretch of grey beach left open by the mangroves, hundreds of fragments of broken crockery were scattered, blue and white and green and maroon, amongst pearly nautilus shells, and old elephant molars, yellowed with age and salt water: all that remained of a living community, already as ancient as millennia's old archaeological ruins. Far in the distance, over the bay, lay a long island, and the Indian Ocean glittered, breaking on the distant coral reef in sprays of white.

We cleared a patch over the sand dunes, and there we pitched our camp. The sound of the waves on the reef was unchanged.

The boat was lowered into the water next day, amongst singing, and *dahos* came over from the island to greet us; fishermen offered gifts of lobster and fish and left in a wake of songs and drums in the evening light.

Time seemed to stand still on the shores of Rubu, facing the long island which changes its names to match the names of the villages along the shore. It was a time of peace and harmony, of exploring the untouched shores where Muslim tradition had sheltered the people from the pollution of so-called progress, and lives still developed following ancient rhythms and customs.

The women were tiny, gracious, slim as girls, wrapped in bright *kangas* and with their hair plaited in smart, tight patterns; their ears were bright with red stones, the slender arms with bracelets, the small hands and feet painted with henna in floral motifs; sometimes a ruby glinted in a nostril, while barefoot children clung to their skirts.

They did not cover their faces, the Bajuni girls, like the Arab women in Lamu or Malindi, nor their heads, and I never saw them wearing a *buibui*, the black flowing, nun-like garment of the strict Muslim ladies of the Coast.

They looked at us from the shadows of their doors, shyly returning our greetings in sing-song voices; and curly-haired little girls stood next to them, enveloped in faded European-style frocks a few sizes too large, but with shiny, eager eyes, devoured by curiosity at our foreign appearance, by the details of our clothes and hair.

I have always been conscious of the scrutiny of those young eyes in Africa, when, at remote villages, the crowd of silent, observing children approaches. Drawn by curiosity they leave the repetitive monotony of the games they have been playing in the dark shade of their homes, assemble at a distance, and finally come close to your car, almost touching it, and there they stop, girls with girls and boys with boys, barefoot, slim, staring silently, absorbing everything, examining the foreign animals we are.

We went for long trips on the boat, circling the islands and waving when we met the fishermen's *dahos* cutting the waves, loaded with biblical figures in long *shukas*, turbans loosened by the monsoon. We slid beside cliffs of steep rocks, hung about with liana and wild aloes and sansevieria, with the odd baobab suddenly appearing silver and majestic through the undergrowth; we spotted awesome sculptures of driftwood on the beaches, carved by the endless fantasy of long waves, and once, a large iguana crouched precariously on the edge of a rock, grasping a branch protruding on the canal, her black tongue flickering, her glassy eyes expressionless. A prehistoric encounter. We watched troupes of vervet monkeys taking over the dusk, jumping with incessant chatter amongst the baobab branches along the shore, where the small, black, left-handed crabs with one giant orange claw scurried down their holes in coordinated shifts, like dancers. We searched for oysters, with an old rusty hammer, and ate them where we were, half-submerged in the tidal ponds, squeezing a lime on their pearly flesh. We went bottom-fishing with Mote as I had once done with Paolo, and caught the same type of little round fish, silver-pink, with a black spot on the side, which Mote roasted for us on the beach on a rudimentary spit cut out of a fresh mangrove twig, and which we ate with our hands, skin and all.

While trailing with our boat in the evenings, we sometimes caught some very large red snapper and Aidan's surprised joy was like a child's, endearing in the experienced man that he is. We cooked them whole, with spices, lime juice and garlic; our camp was lovely, with large, open mess tents like verandahs, and two Lamu beds for sofas, that Mote's brother had made for us on commission, and sold us for far too much.

It was extremely hot, but there blew an evening breeze, the moon rose creating strange shadows, and we were merry.

Over those days of bliss and simplicity, far from the noise and cares of a distant world, the idea slowly formed of finding a place somewhere in that area for us to come and unwind during the busy year, and, in exchange, to start some activity to benefit that community. Something to protect the environment and to give people a way of life.

We sent messages.

So the day came of the historical meeting of the *wazee*, who had travelled from far to hear what we had to propose.

From the beach, from the tracks overgrown with sea grapes, they began to come. They had heard from the fishermen's drums that we had opened again the old road to Rubu, these crazy *wasungu* who travelled by air, but had reached that spot by driving through harsh thorny land, and had pitched their camp on the beach, next to the mangrove bay, below the ruins of the house that had been Mohamed Mussa's.

The news had travelled at night, brought by rhythm and by song. Above the mangrove forests and clumps of doum-palms, through the tall tamarind trees and the outcrops of coral rocks overgrown with green creepers, to the pockets of people still living in clusters from Kiwayu to Kiunga, as far as Mkowe, through the fishermen's camps on the island, people listened and learned of this strange and unusual fact.

They knew that we travelled with Mote, *mzee* Mote from Mkokoni, and that we had hired Haji's son Ali. Some of them had already heard, when they were consulted for the spelling in Arabic of the word 'Rubu', painted in ocean-blue on the white side of the boat. And from Mkokoni to Kiwayu village they all could see Mote beaming with anticipation and growing in importance in his own eyes because of the association with us. From his behaviour alone they had guessed some excitement and novelty lay ahead. Who were these *wasungu*?

Everyone knows that *wasungu* are slightly touched: those who can sleep in beds and sit comfortably inside, but choose incomprehensible hardship, like walking for miles in the sun following camels, without being obliged to do so, or who sleep out in the open under a mosquito net in lion-infested areas, and come all the way to this god-forsaken land, remote from man and civilization, not a *duka* for hundreds of miles, not even a road to get there. And what for?

I presume it was an irresistible curiosity to find out why, and the concern that we might have some sort of plot for Rubu: they wanted to claim the place for theirs, as it was; to show that they had first stake, that they cared what went on there, even if for thirty years none of them had taken up a *panga,* called for companions and, slashing at the vegetation, for about fifteen miles, made their way through tsetse country on the old road to Rubu. Now the track had been opened and the spell was broken.

Speculations of all sorts were born each morning and died with each sunset, to be reborn at sunrise and discussed all day long. But now they had gathered to hear from us directly.

Alerted to the arrival of the wise men, Aidan had driven to Kiunga earlier, to fetch Chief Jamal, the DO and the assistant chief.

I served everyone *bajuni* tea, and biscuits, under the awnings of blue and turquoise *kangas,* and then they all gathered under the tamarind tree below the *jabia.* After a moment's hesitation, I declined Aidan's invitation to join them because this was a Muslim world and my presence at the start would have made them uneasy: I knew women did not appear unless called. I felt that, by this show of instinctive respect of their tradition, I would myself gain in respect, and this was auspicious. As it happened, shortly after the meeting began they sent for me, and it was a prouder thing to join them at their request than to sit there uninvited at a meeting in their own land.

The meeting began with prayers to Allah. They then spoke formally in turn, welcoming us amongst them, asking what they could do for us. They had been impressed by our determination to reach that spot, and at our respectful way of proceeding; they had learned that we were looking at the land and wanted to know what we had in mind. There were some old men, and some others quite young, whose wisdom had given them seniority. They sat cross-legged on the sand, dressed in clean long *kikois* and shirts, all wearing embroidered Muslim caps, apart from the DO who was a Christian from Nandi, Mr Towett Maritim. He was still quite young, and very agreeable. Slim with an open, intelligent face, he was gifted with a positive attitude and natural leadership. He spoke very well, in excellent Swahili as they all do, in a sing-song of coastal accent, very pleasant to hear, the melodious intonation of all fishermen dialects across the world: a tone and a *cantilena* pitched to cover and pierce the sound of waves, to call from boat to boat over the oceans.

We told them of our love for that land, and of my memories of long ago; we explained our busy lives and our many commitments, but also our desire to find a place of quiet and peace, and of our intention – should they agree with our offer – to help them to rebuild and develop their village in a way that respected their old customs and their environment.

I looked at Aidan over the crowd of elders: another commitment, I could read in his eyes, with all the thousand things that already crammed our days. Yes, *wasungu* are mad, why ever start all again from the beginning, pioneering far from our known lifestyle, and our comforts?

But they listened with deep interest. We represented hope and change, adventure and challenge. Their eyes lit up, Allah was again invoked, they shook our hands repeatedly, they asked us to look around and to give them our proposed terms.

It took quite a while – like all things of worth – subsequent visits, letters, meetings in darkened offices in the old Lamu town, under palm-trees in the dusty town of Kiunga, in banks with fans whir-ring from ceilings of peeling plaster, and in open fields of cassava, where we managed to land to meet one more of the old people and get his blessing, too: the unavoidable bureaucracy which is the other face of most dreams.

But finally, it was nearly done. Through my enthusiasm and Aidan's perseverance, his knowledge of the rules and his fairness of play.

It was for me the crowning of a dream long shared with Paolo, but which there had been no time to accomplish, to have the use of some land at the coast, and some sort of dwelling there, built to please the eye. Somewhere where the noise of the waves on the reef was the last sound at night and the first in the morning, a rhythmic, heaving sound like the beating of a large heart.

The image of Paolo kept coming back in that place he had loved most. Then we would sit and listen for the roar of lions, in the nights of those two lost weeks. And above the noise of the waves on the reef, brought by the *Kaskazi* wind from the sandy acacia-dotted savannah below the baobabs, sometimes we heard, with a shiver, the powerful deep call which is so much the sound of Africa.

Memories glided back with the tides at Rubu, when I paused on the shore, alone, to think, like waves reaching out on the grey sand between the mangrove sprouts.

Memories of Paolo.

Buffalo had enthralled him above all other African animals.

There, to his emotion, we watched a few old males coming down one evening to wet their muzzles in the semi-sweet brackish water amongst the mangrove.

Paolo.

Where was the shadow on the dunes, the tall slender shadow of the man who has gone? Why could we not say our farewells? The long wait for him to come home, his voice on the telephone, disembodied, far away, then the shade growing longer on the lawn, a gloom descending with dusk, the intuition of loss, the ultimate pain, the startling agony of bewildered, foreseen loneliness, the knowledge of nevermore. *I put my hands around my stomach, felt the life I was carrying, knew the other life had gone on. I knew there would be no more holding hands, eyes gleaming with life and excitement, deeply searching mine in a quest of love.*

'Farewell' – I thought – 'my Paolo, Paolo of the golden hair, Paolo the chivalrous squire.'

And Paolo had left when his time was up. His time was up and there was nothing anyone could do. Circumstances, delays, a variety of events, brought Paolo to that spot where, at the same time, the faceless lorry-driver chosen by destiny decided to turn into Hunter's Lodge. And it was at Hunter's Lodge that Paolo the hunter met the trailer, when his time was up.

I knew it with a jolt which left me breathless. An abrupt intake of breath, as if his spirit, abruptly freed from the cage of its body, had exploded with the same powerful energy of when he was physically there.

And Paolo touched my face with his fingerless hands, caressed my stomach where his hidden child curled, slowly growing into the blonde, radiant Paolo-girl he would never see. The one who was now running towards me on the beach, long-limbed, suntanned, laughing.

Memories.

Memories of running uphill, sitting on windswept rocks at the moment of sunset, when there is a chill in the breeze, shivers run from the shoulders down, and we look at the lake, at the blue Cherengani hills, at clouds with strange shapes, like dragons or fish. Of running downhill, hand in hand, eyes in the eyes, tripping on stones, and laughing.

Memories of long nights of music, the fire blazing, talking and talking.

Of other dogs sleeping on carpets, of other flowers fading away in vases.

Memories of brown, warm eyes, intelligent and old, even if young. A solitude, a far-away wisdom.

My lost child, my lost boy child.

In the months following our first visit we returned to explore with new eyes, no longer passing visitors who might never return; eager to find a good landing strip, slowly driving our boat up through the sleepy mangrove-lined canals at high tide. We discovered it at last, at a corner of the canal bordered by banks of red earth, below a hill crowned with baobabs and filigreed with porcupine tracks, the perfect landing strip. Bleached mangroves, like ancient noble bones exposed to salt and sun, lined the sides, and there Aidan landed his plane with ease.

It was late afternoon, almost sunset. Just enough time before darkness fell to pitch our tent beside the plane, gather a pile of firewood and light a fire to bake bananas and to boil water for tea. Fireflies danced in the warm night and from the coastal forest came the calls of bushbabies dancing high in the baobabs.

The following morning, leaving the plane on the shore, our tent and equipment a few metres from it, we jumped into our waiting boat. Mote left us on the beach, at Ndoa, and went to have his *chai* at the fishermen's camp on the island. We walked inland, exploring, following tracks of bush-pigs, and to our amazement saw a male bush buck jumping, alert and handsome, through the tangle of sea vines.

We went down to the sea to swim and, in a grotto left open for us by the receding tide, we set our picnic.

We were eating cold fish with lime and soya, when we saw Mote sprinting towards us. He leaped over the rocks, agile and lithe like a sand crab, and he was screaming. The ocean wind took away his words, and only when he was close we heard:

'*Moshi! Ni moshi mingi kwa campi yetu. Moto! Ni moto kwbwa . . . Ndege . . .*' 'Smoke! A big smoke in our camp. Fire! A large fire . . . the plane . . .'

We already were running before he had finished. Gathering our gear, wading into the water up to our armpits, climbing into the boat, firing the engine. With a wide circle the boat gained the entrance of the *melango*, left the ocean boiling behind, reached the lagoon where the horizon spread before us. Standing on the rolling boat and sheltering our eyes with our hands, we watched in silence.

High into the sky, lifting exactly from the place we knew as Abukari – where, hidden behind the mass of green leaves, we had pitched our camp and our aeroplane was parked with all we had brought – over the mangrove forest, moving into the wind, ugly with the inevitable, lead-grey colour of fear, was an enormous column of smoke. No one spoke.

In the distance, running up and down the short beach at Rubu, we could see the small figure of Mohamed Mussa gesticulating towards us, urging us to hurry. The tide was out, the water low, exposing banks of white sand, and there was no way our boat could reach the bend in the canal and the airstrip. We had to walk.

We jumped into the water before reaching the beach and dashed off. Mohamed led the way. We flew uphill over the sand dunes and reached the swamp. The forest closed over our heads. The mangrove roots stood out, caked in mud, like gigantic spiders. I tripped a few times, losing my sandals. The sucking noise of crabs and raucous calls of parrots, a fish eagle taking off like hope from the treetops into a perfect sky, our steps deep in the gluey sand, the beating of my heart, my fear, my prayer.

Yes, I did pray to those who have always helped. Who have gone ahead where time means nothing, where anything can happen. I prayed with an intensity which left a dry salty taste in my mouth, tears stinging my eyes.

I prayed for the impossible to have occurred; for the random, compassionate breath of Providence to defeat the chronology of events, and to blow away, retrospectively, the destructive fire; for the clouds to melt into rain and soak the embers.

All the time I was thinking of the aeroplane, of the tanks full of aviation fuel we had left below its wing, of our breakfast fire that had seemed too close, that we had believed extinguished; of the unbearable heat that the fire must have caused . . . was it my imagination or was a blade of fine rain falling ahead, obliquely . . . was it my imagination, or was the smoke getting lighter? No longer like a leaden column, but drifting wispily in the breeze?

The last mile was the longest. We reached the canal and plunged in, partly swimming, partly wading up to our armpits. Then we came past the bend, and we saw it.

Up to our waists in the receding tide, Aidan and Mote, Mohamed, Ali and I watched, speechless.

Along the middle of the long strip of sand fringed with mangroves, several metres wide, half a mile long and beginning at the water's edge, was a dead stretch of burned grass, of charcoal stumps still glowing, like a grey carpet unrolled by that inferno. All the firewood we had accumulated below a thorn tree had been destroyed. A scorching, choking stench of smoke remained.

No more than two inches away from the smouldering ashes, on opposite sides, incongruous, intact as we had left them, were the aeroplane and our tent. Only one of the tent's strings was scorched and hung limp. In the distance, smoke still rose, some shrubs still glowing, but, inexplicably, the fire had subsided on its own.

Mote was the first to climb up the beach, where he fell to his knees, arms outstretched, praising Allah. In his way he voiced our own feelings: to the day we die none of us will forget that that morning at Rubu we had witnessed what we could only call a miracle.

I looked around at the baobabs, at the bush, at the sparkling sky. I looked down at the sand as I stepped from the water. It felt right and it was like an omen. There was a large lion track, fresh from the previous night, and a buffalo, I could see, had come down to the sea shore.

The Ring and the Lake

Or speak to the earth, and it shall teach thee.

The Book of Job, XII, 8

There is in Colorado a very special place called The Baca, an immense ranch spreading at the foot of the Sangre de Cristo Mountains, where the air is thin and heady, the sunsets long and deep red, and the light golden and pure as I have only found it before in the Highlands of Kenya.

In the mornings deer come down from the snowy mountain they call The Mother, and you can see them grazing undisturbed, like impala, on the bleached grasslands which remind me of the savannah. The flowers are yellow and blue and have a dry, long-lasting quality. A wild sage, valued by the Red Indians for their ceremonies, grows on the mountain slopes, and its aroma is the same as that of the *lelechwa*, the sage which grows wild in Laikipia on the edge of the Great Rift Valley.

Like Laikipia, from where you can see Baringo, one of the many lakes for which the Rift Valley is famous, The Baca opens up into the great Saint Louis Valley, which was once upon a time an immense lake, now sunken out of view. You can still sense there the proximity of water in the smell of the wind, in the vegetation – pines and cactus and herbs of an almost Mediterranean quality, in the parched sandy terrain which reminds you of the shores of an African lake, and in the astonishing miracle of

the Great Sand Dunes, a small ever-changing Sahara of breathtaking beauty, whose shivering patterns are like those left by the
waves of a receding tide.

The place has long been renowned for its spiritual power, and
since time immemorial was used by the American Indians of all
tribes as a healing ground where, their wars forgotten, they could
come to bury their hatchets and worship their gods, dedicating
themselves to the sacred ceremonies of their traditions.

It was there that I met Sheelah. She was an Indian priestess of
an ancient Vedic sect, one of a number of religious groups who
had been welcomed to The Baca by its enlightened new owners,
so that the place would continue to be a spiritual retreat on a
larger scale, embracing all of the world's old religions. It was a
time of soul-searching for me, after the death of my husband and
son. With my small daughter, Sveva, I had accepted the invitation
of Maurice Strong and his wife, Hanne, who, after visiting Laikipia, had seen the extraordinary similarities between the two
places. They suggested that I would find at The Baca something
which would help to heal my wounds.

'There is a fire ceremony tomorrow morning at seven down
by the creek. Come,' said Hanne, the first night we were there.

A fire was lit every night at my graves in Laikipia, at the bottom of my garden. I found fire evocative, purifying, and I was
intrigued by the idea of a fire ceremony. We went.

We followed a shady path along a stream, beside the gentle
murmur of running waters, to a glade where a group was assembled around a large fires built in a pit in the ground. Symbolic
offerings of fruit, flowers, rice and honey were assembled on one
side, a large shining rock crystal reflected the morning light, and
fine incense smoke drifted in the cold, clean air. People were
sitting in a circle, and among them, dressed in a red sari, was a
small woman.

Instantly I had the odd feeling of having met her before; and when she turned to look straight at me and smiled, I felt I had known her forever.

'I am Sheelah Devi Singh. Welcome to our fire ceremony.'

Her eyes were brown and simmering with a bright, hot light. She took my hand, and her fingers were hot and dry and burning too. An arrow with a red tip was painted on her forehead, like a flame; the fire cast orange reflections on her olive skin and short cropped white hair; my feeling about Sheelah, from the beginning, was that she was made of the substance of fire.

It was a simple ritual with mantras and ancient songs in Sanskrit, to thank Mother Earth for her gifts, to give her back symbolic offerings and to ask for simple wishes of peace and healing to be bestowed upon us. I found it soothing and timeless, and I gained from it a sensation of lasting harmony, a quiet in my soul. It was right to give back some of what we took. I have learned since, in fact, that what one gives always comes back, often in various forms and different ways.

I saw Sheelah often after this, during subsequent visits to The Baca over the years, and occasionally attended her fire ceremonies in the early mornings. She belonged to the Rajput tribe, the most noble of all, whose warriors ride into battle carrying spears. She had followed her religious calling after an unusual life, and, although my rational and independent mind has never allowed me to adopt a specific faith, nor to give my spiritual search a definite label, I found her philosophy and her music a soothing and positive presence at that time of my life.

A bond older than friendship seemed to grow between us. With her I went through the unforgettable experience of transcendental meditation through deep breathing, which lifted me out of my own body and made me go back into a forgotten past, where images of former lives disclosed themselves to me, and left me with an accomplished feeling of great peace, of total happiness.

Indian beliefs had always fascinated me. Sheelah explained

them to me with a mystic simplicity which I found ineffable, and when she sang with her harmonium, the sounds of the unknown words dug in my soul odd echoes of reminiscence.

When Sveva was about to be eight years old, we went to The Baca for a few weeks in the summer. Unknown to us, Sheelah travelled all the way from Bombay, where she now lived, to be there for her birthday. Eight is an auspicious number in Eastern tradition.

Sveva would be eight on the eighteenth day of the eighth month of 1988; we were at 8,000 feet; there were many people, and when we counted them it did not seem like a coincidence that they were eighty-eight. I gave Sveva eight presents, and the last was a magic wand.

Sheelah had brought gifts of old silver, incense and silk, and Hanne had arranged for an old hippie from Boulder to come and play his cymbal to Sveva for good omen. Its vibrations reverberating in the morning air sent shivers down our spines.

At a special fire ceremony for her birthday that morning Sveva, dressed in red, with a flower behind her ear, fair hair streaming down her back, was arrestingly beautiful. She took all this with a serious grace that was exquisitely hers, and I felt that allowing her the opportunity of experiencing, while still so young, many aspects of the human search for the infinite, could only add to her inner growth.

On the day before we left to come back to Africa, Sheelah unexpectedly knelt in front of me and took from her own feet the two gold rings which adorned her middle toes. Without a word she put them on mine. In her brown eyes was a loving warmth mixed with an inscrutable sibylline detachment. Her voice had a sing-song accent and an almost hypnotic quality.

'You will wear these as if you were a Rajput, and you will never lose them. You must have courage. You are my sister, and we shall meet again, perhaps in other lifetimes. You will travel the world, and many people will be around you. You are going to

write a book, and the next few years are going to be full of action.
You will succeed in your dream through your hard work. It will
be up to you, and you will make it. Remember that the choice is
always yours. Never give up. You have much to do. Emanuele is
at peace now, and Paolo . . .'

She looked at Sveva, who smiled at her and at me with Paolo's
eyes.

I wore my rings on my toes always, after that day. When people remarked on them, I told them it was a long story.

I wrote my book, and this kept me away from The Baca for
some years. The Gallmann Memorial Foundation grew, and its
activities took all my time, all my attention. I wrote at night, like
an owl, so as not to steal time from my daily work. I established
the education project in Emanuele's memory and I developed
many others, too, in Laikipia. Aidan flew back into my life, to
bring passion and adventure. Sveva grew harmoniously, and I felt
fulfilled.

I never saw Sheelah again. A few years later I heard that she
had died as a result of falling from her horse at The Baca. I felt
her loss, but I knew that it was a fit way to move ahead, for a
proud Rajput.

The rings became even more precious. Occasionally it happened that one or the other got caught in the grass, or in the
thick fabric of a carpet when I was walking barefoot, or among
my blankets; but uncannily they were always found again. So
much so that I came really to believe that I could never lose them,
and this became a joke among my house staff in Laikipia.

'*Pete yangu ulipotea tena.*' 'I have lost my ring again,' I would
tell Julius.

'*Sisi tapata, tu. Wewe awesi kupotea hio pete kamili.*' 'We shall
find it. You cannot really lose that ring for good,' he would
answer with a smile.

And, sure enough, a few hours later he, or Simon, or Rachel, or one of the gardeners, would appear with my toe-ring in their hand.

Sheelah's rings became a sort of special talisman, and looking at them glinting on my toes I never failed to feel pride, gratitude and a certain comfort.

Last summer Sveva, Aidan and I flew to Lake Tanganyika. We left the plane on a tiny airstrip cut out of the tropical bush, next to a village, built only with natural materials; plastic, tin and cement had not yet reached the lake shores. It was an amazing place, still belonging to a remote time that so-called civilization had not managed to alter. On a peninsula outside the village we saw, while passing in the boat bound for our camp, a weird scene of witchcraft, as if in an account of Burton's early explorations. Seven dead cats were hanging on shrubs to propitiate the spirits of the water.

Extraordinary trees grew along the lake, and tangles of rare plants to Aidan's delight, and the time was one of bliss, love and sheer happiness.

We stayed in a fantastic camp of white tents, like a sultan's, that stood on a white beach, where we were the only guests. It was a place of pure enchantment. Every day we walked up into the forest to look for elusive chimpanzees. We swam in the cool water and went for sunset expeditions in the boat, fishing in unvisited rivers for small yellow and black fish that appeared surprised to see us.

On the last day we went for a long walk up the mountain, alive with butterflies and strange creepers and liana. It was extremely hot, and after trekking up and down the hills we welcomed the freshness of the lake water, which is as transparent as the most crystal-clear sea. We took off our shoes and our clothes, left them in the spreading shade of a large mango tree, and ran into the lake to swim.

It was while walking back to the camp along the shore, our feet in the waves, that I realized one of my rings was missing. The currents were powerful, with a swell that was curiously similar to the ocean tides. The sand was coarse, with small shells and pebbles, and it sloped steeply down below the water. My feet had sunk into it, and there was no way a small, non-floating object could ever be found again once it had been captured by the sucking waters. It looked as if this time my ring was lost forever. I felt deeply sad, deprived.

We walked up and down the beach many times, searching in vain amongst the debris for a golden glint: but we knew that there was no point. Turbulent waves continuously swept the shores clean. The proverbial needle in the haystack would have been infinitely easier to find than my toe-ring in the lake. A white depression on my toe marked the ring's place. It would fade in time, as I should not substitute it with another.

I was crestfallen for part of the day, but then, as the shadows grew long, I became resigned to this irreversible loss and understood that it was up to me to accept it, and let go. I decided that I should make a special, positive gift of it to Lake Tanganyika in Sheelah's memory. In her fire ceremony she spoke to the earth and gave it back symbols of what humans thanklessly took: plants, water, scent, minerals. A golden ring was the perfect offering.

At sunset I went with Sveva to the waterside. We both noticed that, in an eerie way, the colours in the sky were the same deep blood red as the Colorado sunsets. We recited ceremonially an almost forgotten mantra that Sheelah had taught us in the old days, the mantra of giving, which ends with '*Swaha*', the word which was pronounced after each offering.

I thought vaguely of the Venetian Doges who every year threw a precious ring into the Lagoon, in a symbolic marriage with the sea.

Now with a light heart I offered to the great lake the ring that it had already taken; with thanks for its beauty and the happiness we had experienced there.

'Lake Tanganyika, I offer you my ring. I am glad that, if it had to go, you took it. *Swaha.*'

I felt somehow relieved after this, as I knew it was the right thing to have done. Hand in hand, I walked with Sveva back to Aidan and our tents.

We had finished packing, the next morning, and the boat was waiting to take us back to the airstrip carved out of the forest. I was in my tent having a last look round before going.

The man came running from the mess tent and stopped a few feet from me.

'*Memsaab,*' he said, '*nafikiri hio ni yako.*' 'I think this is yours.' '*Ulikwa kwa mchanga.*' 'It was on the sand.' He held out something in his hand. It glittered, and in the morning sun it seemed to wink. Behind him, the lake was glimmering, benign and generous, with all its secrets.

With a thumping heart, while Aidan and Sveva watched in total silence, I reached out to take back my ring.

A Dance of Spiders

It was as though, in that remote corner of the world, I had
received a sinister hint as to the existence of certain
influences outside the palpable terrestrial sequences of life.

LLEWELYN POWYS, *Black Laughter*

A tragic thing has happened. I am terribly sorry.' The voice
was serious, strangely hesitant. I tightened my grip on the
telephone receiver.

A silence, charged with premonition of doom.

'It's Julius.'

The red-haired man swallowed another sip, and his restless,
intelligent eyes became misty behind his thick glasses. I was glad
I had fixed him a very strong whisky: it looked as if he would
need every drop.

In the candlelight of the sitting room of my Nairobi house, his
face was lustrous with sweat. His eyelids were red-rimmed, and
there was a slight tremor in the hand which held the glass. His
voice had an inspired intonation, the emphatic quality of Sunday
sermons, the sternness of Dickensian teachers in young boys'
schools.

'Talk,' I encouraged him. 'I think it would help you if you
could tell me the story from the beginning. How did you meet
Julius?'

His freckled face lit up in a smile.

'I will never forget the first time I saw him. Many years ago, when I had just begun teaching at Amani Boys School, I went with a colleague to a Kikuyu village looking for candidates for a scholarship to the school. He stood out amongst all those boys with a presence that struck me instantly. The peasants listening to him hung silently on his lips. Still so small, he was aloof, unreachable. From that moment I loved him.'

A glazed look clouded his eyes.

'He was nine years old.'

The early afternoon sun brightened the fields of coffee bushes and the red wet earth trampled by many feet. Banana trees, planted in clumps, shaded rows of young maize shoots. A mottled group of cattle, goats and a few hispid, mud-spattered sheep grazed off the long elephant grass that grew green and abundant on the slopes of Kiambu. Grey clouds, just emerging from the horizon, announced the daily evening storm of the short rains. From the country church, a melodious sound flew in waves, young voices singing a lilting hymn that came from distant lands.

Standing in the churchyard, facing an audience of villagers of all ages and both sexes, was the church choir of young boys. They wore long robes of vivid blue, with wide white collars falling over their shoulders. Their eyes in the dark faces were bright, imbued with a strange sense of wonder, and they sang with the rhythm and enthusiasm of innocence and unquestioning faith. Their features were still childish, unformed, smooth, the voices pure and high.

He stood out amongst them. He was not the tallest, and his harmonious voice was not the strongest.

Perhaps it was his beauty. The clear brow, the curling, girlish eyelashes, his smile. It was, perhaps, the way he held himself, radiating pride and confidence.

Perhaps it was the feeling that he was remote and difficult to conquer.

When the singing stopped the boy came forward a few steps,

and walked out on to the patch of grass before mounting into the wooden pulpit. His voice was frail and crystalline, yet enhanced by a quaintly powerful authority. He addressed the colourful crowd with his religious exhortations and a spontaneous grace which went beyond prayer.

Anthony looked around, observing his fellow spectators. Old women, their grey heads in scarves, dressed in long skirts and sweaters hand-knitted in bright yarns; young women, carrying invisible sleeping infants well wrapped in pink and green lacy shawls tied to their backs; old men, young men, heads bare, still mostly in their shapeless working clothes. The lazy afternoon insects buzzed in the sun. No one moved. The crowd was spell-bound, enraptured by the boy's charm. Standing in the half shade of an avocado tree, Anthony watched to the end. Then he approached.

'Would you like to study in a big school in Nairobi, where there is a great choir, and music matters?'

The boy looked up: large, liquid eyes with impossibly long, silky eyelashes. The boy smiled, a dazzling flash of even teeth like matching macadamia nuts, a warmth of honeycomb in the hazel eyes, which seemed to touch and melt something in Anthony's chest, and his legs were reduced to wax.

'Yes, I would.'

'Do you think your father will let you?'

'I have no father. He died two years ago.' A shadow drifted across his brow like a curtain which was soon lifted again. 'He drank. I have four brothers. I am the eldest.'

Anthony swallowed.

'And your mother?'

The little boy shrugged.

'She would not mind. She has much to do. There is no money and she works in the *shamba*. She also helps in the church.' His face hardened and his eyes clouded over. 'After church, she goes to the *muganga*.'

'I could look after you. I teach there. Would you like that?'

'Perhaps,' said the boy with an adult smile, 'you will let me read your books? I like to learn about animals and nature. I have no books of my own.'

Again the sudden white of even teeth.

'My name is Julius.'

My name is Julius.'

The handsome face smiled at me and I noticed the eyes which seemed to take over most of his face.

'A good-looking young man,' I thought. He wore a pair of clean new jeans, a striped blue and white cotton shirt. A plastic watch, new and brightly coloured, was fastened round his slim wrist. I looked at his feet: good leather moccasins, a blue pullover slung around his shoulders. His skin was an amber brown, unblemished. Short curly hair and fine hands with clean short nails. On his upper lip a shadow of a moustache. He was slim, with narrow hips. Slightly effeminate? There was a distinctive radiance about him which could not be missed.

'It is very kind of you to ask me to dinner,' he said carefully in good English. He looked up at my tall *makuti* roof, around at the polished furniture, the gleaming copper and brass vases. As on every night, a candle trembled in front of the silver frame with Emanuele's photograph. A fine blue smoke drifted from the incense holder. My eight dogs slept on the carpets in various poses of abandoned contentment. Two servants in starched white uniforms were preparing the table, laying red hibiscus flowers. The fire crackled vivaciously in the vast fireplace.

There was a candid wonder in his eyes.

'I have never been in a place like this before.'

A few weeks earlier I had received a letter, signed by Anthony S., his teacher, sponsor and protector, who asked if we could host in our research camp one of his most brilliant students, the head prefect of Amani, who needed first-hand experience in the outdoor environment in order to write a descriptive essay which would allow him to run for a scholarship to Cambridge University.

Amani was well known for its high standards of education and for making available superior academic training to students of underprivileged background but with good minds; it was known to instil discipline and a sense of responsibility. In the firmament of Kenyan schools, Amani was a shining star, unique. There were well over a thousand students there. Whoever managed to become head prefect must have been rather unusually gifted with leadership and intelligence. But even so, Cambridge University was still as far from Amani as a distant planet.

I had been intrigued and had immediately agreed to host this boy for a month or so.

Looking at him now, I felt that I wanted to help his ambition to materialize, and that I would see more and more of him over the years. There was something special about him, which I could detect immediately, but not describe.

'I understand that you would like to read geography at Cambridge. A really ambitious plan.'

Not for a moment was there hesitation in his tone:

'I have been fortunate. I want to give back to my country what I have received over the years. Cambridge. I am sure I will enjoy it there.'

I opened my bar cabinet.

'Would you like a glass of wine?'

For a moment only, he looked lost, then he grinned.

'I would rather have a Coke, if it is all right with you. I am not used to wine.'

We can't find Julius.'

Anthony's voice sounded tight and strangled on the phone. 'He has disappeared. Gone. I can't locate him anywhere. He is

meant to leave for Cambridge in two days' time. I was hoping
you might have heard from him.'

His emphatic preacher's tone became a whine.

'What are we going to do?'

I answered with the calm that always prevails in me when others around me lose theirs.

'Take it easy. He is practically a man, remember? Why should
he always report where he goes? He has to be able to cope on his
own in England. I am sure there is a perfectly ordinary explanation. Do you have any reason to think otherwise? Perhaps he has
gone to say his farewells to some friends.'

'That is exactly what I fear,' Anthony whispered.

The noise of a car engine was followed by the slamming of a
door, then of another. Footsteps descended the stairs to my front
door. Someone knocked. I went to open.

Julius stood forlorn on my doorstep, in a dejected attitude,
head bent and the collar of his wind jacket open on his thin neck.
The change from the secure, radiant boy who looked me always
straight in the eyes struck me like a blow.

'Julius. Are you all right?' No answer.

'Julius, for God's sake. Come on. Come in. Tell me what's
happened. It can't be as bad as that.'

A movement behind Julius, the leaves of the gardenia bush
swayed. From the shadows, another shape emerged. The sudden-
ness of the movement startled me. I caught my breath.

'Get in,' said Anthony, as if talking to a rebellious little boy
who had thrown a tantrum. I noticed his eyes were enlarged. He
pushed him in.

'You must confess to Kuki what has happened.'

Julius sat on my white sofa in the soft light like a culprit ready
to confess a crime. The absurdity of the situation annoyed me.

'What is it?' I probed gently. 'There is nothing at all that you "have" to tell me. Your life is your life. You look fine to me. Can I help you in any way?'

'Tell her,' urged Anthony.

I braced myself for the worst, and yet I felt there was really nothing the boy could have done so tragic as to justify such attitude.

'A few years ago I met a girl,' murmured Julius in a small, dull voice. 'She was one from my native village. I have known her all my life.'

'And so?' A lightness overcame me, as if the precipice on which I was perching had revealed itself much shallower, less forbidding than I had feared.

'You must tell her. Confess what you have done.' Anthony's tone sounded absurdly serious.

'I have a son. He is three years old.'

An immense relief swept over me.

'Is that all?'

Julius looked up, and his moist eyes held an appeal. But through this, to my dismay, I read an overwhelming terror, as of a haunted animal running blindly off in the night.

'I have received this.'

He offered a crumpled sheet of paper, which blossomed in his open palm slowly, unfolding like a crushed moon flower. Words crawled over it, scribbled in pencil in gangly, uneven letters, oddly repulsive, like a dance of spiders.

'I can't read Kikuyu.'

'It is a threat.' His voice broke in a sob. 'They have put a spell on me. They do not want me to leave. If I go, I will die.'

A chill drifted through my western room, ugly with sneering masks and unmentionable rituals, slaughtered black goats, smeared entrails on muddied feathers, beating night drums, guttural songs and young boys' fears. For only a moment its power hit me, the unease prevailed. For that moment I could believe anything, I accepted that anything could happen. Once again I knew that this world I had chosen was not just red sunsets, migrating herds, sunny savannah and bougainvillaea in bloom: I remembered the snake, the screeching of imagined brakes, the open graves, the pain, the ephemeral reality which often eluded our comprehension and mocked our bewilderment: and I realized that this, also, was Africa.

I came back to the present.

'Julius, this is 1985. You are about to go to one of the most sophisticated universities in the world. You are going to fly there in a large aeroplane, meet many people from all over the world, learn useful things which will, when you come back to Kenya, allow you a career that is all you have ever dreamed of. By going you are hurting no one. This is what the new Kenya is all about. They should be proud of you, back in your village.'

A thought raced quickly through my mind and I chased it off.

'Who is it? Who has written this? Who has put a spell on you?'

His face had closed like a slab of stone, and I could see I would not learn it from him. Not now. Even as I was asking I knew that he would not tell.

'They think that I will not come back. They think I will be out of their influence.'

'The girl's family? Do you care for this girl?'

Julius had no time to answer.

Anthony looked up sharply with pressed lips.

'He does not see that girl any more. It was just a mistake. It was her fault. The girl has given up any hope about him. An ignorant peasant girl. She was compensated.' He waved the shadow of the girl aside like an insignificant moth.

I reacted.

'However, her family may think otherwise.'

Julius looked up at me with a sudden resolution in his eyes. I read confusion, fear, hatred and resignation all at once in the look he gave me. I read something else also, deep and murky, unspeakable and terrifying, which I could not interpret.

I cleared my throat.

'You must go ahead and nothing is going to happen to you. Nobody can hate you just because you go to school to learn. Spells and *muganga* cannot hurt you if you do not believe in them.'

With immense horror, I realized he did.

A shiver of rain forest and dark huts, green smoke and voodoo passed again over the glittering crystal table. As in a Gothic tale, in my imagination the candles flickered uncertainly.

The sound was low and hoarse, it was his voice no longer. It rustled like a snake and I recoiled.

'The girl did not go to the *muganga*. It was my mother.'

Three years later.

I closed the heavy door, the chill of November behind me. The bright light caught me unprepared, blinding me for a moment. The vast room was full of people and the noise of conversation echoed from the high vaulted ceiling.

I focused on the unreal scene with wonder.

The elephants moved in their cages set in a circle, swaying their sad grey heads rhythmically from left to right and from right to left.

Waiters in white jackets and white gloves circulated through the guests offering trays of canapés and cocktails. The air was warm, humid; candles glowed trembling in silver chandeliers. I looked around, recognizing some familiar faces, and then I saw him.

Holding his glass of wine, still intact, dressed in a blue suit and starched shirt, a very good silk tie impeccably knotted, Julius smiled his disarming warm smile at me across the room.

An elephant trumpeted, interrupting all chatter of conversation for a few moments. The whiff of steaming dung could not be disguised by the mixture of perfumes and cigarette smoke. It was a very weird setting for a cocktail party, the Elephant House at London Zoo.

Julius had been asked as my guest.

'Thank you for inviting me. I feel somewhat overwhelmed.'

So did I for that matter. The absurdity of the occasion stood between us like a pond to cross, and I smiled.

'These poor creatures. And all the real ones, the free ones that we have left back home.'

He scowled, the corners of his mouth turned down, betraying an undisguised nostalgia. For an instant the sadness of the exiled elephant was his same sadness.

'Anthony tells me that you took to Cambridge like a duck to water.'

He looked groomed and confident enough, very much in command in what truly was an unusual environment.

Despite – or possibly because of – the trapped miserable elephant, it was a long step from Africa.

He grimaced, waving his full glass gingerly. 'I like it there; but I still cannot drink the wine.'

I noticed that his fingers held the stem very tightly, with white knuckles. For a moment I thought he was going to snap it. A black crocodile-skin strap circled his thin wrist, setting off a new gold watch.

'Are you sure you are all right, Julius? Are you looked after? Are you very homesick?'

Such a long step, I thought, from those mud huts, the thatched grass roofs, the banana clumps and the smiling girls with high breasts and singing eyes. Such a long step from his real roots. If I felt somehow out of place, how would he feel? Social superficial chatter, witty whimsical remarks, politics, the economy, the last show, what happened to those rhino in Meru, the price of ivory, absurd about rhino horn trade, is Leakey's appointment really going to make a difference to Kenya's wildlife?

A lady in a black velvet suit, a blonde lock falling artily over one eye, took hold of Julius's elbow, swept him away effortlessly and in seconds he was lost in the crowd.

For a moment I saw him as a shipwrecked sailor taken away by the tide, never to be reached; I felt I had to rescue him; then someone attracted our attention by chiming on a crystal glass, and the speeches began. When I looked again for him, he was gone.

Julius, Julius, how nice to have you back!'

He had grown a light moustache.

'I passed the exams at Cambridge. Now I think I will go for a Ph.D.'

He seemed thinner, taller. His accent was more polished, the words chosen slowly and with care. He was detached even when answering. I felt his laughter as a superficial way of hiding a deeper concern. There were corners of his personality where he did not allow anyone to probe; what did he have to hide? What was he afraid of? What tangled cobwebs hid behind the sunny boy's mask?

'Anthony came to fetch me at the airport.'

'Nice of him, at that hour. How is Anthony?'

He lifted his shoulder with a distant look.

'I had not seen him in months. I have a girlfriend.'

He looked me straight in the eye.

'He does not approve.'

And added, in a hiss that startled me by its concentrated spite: 'As if I cared.'

'A girlfriend, but that is great, Julius! Who is she?'

'Her name is Olinda. We live together now. I love her very much.'

To my surprise, his eyes filled suddenly with tears:

'She is so beautiful.' He whispered it with a longing in his voice, as if talking of a fairy which eluded him.

'You must be happy, we must celebrate. It is your life. You are in your twenties. It is perfectly normal for you to fall in love. I think it is wonderful. Anthony must accept this, of course.'

What had never been said struck me, and the time was right to finally bring it into the open.

'He loves you too,' I added gently.

'I know. But that is not the problem. I can deal with Anthony.' A far-away look. 'I have done it all my life.'

'Why are you so sad then? What is there to disturb you? Perhaps you can tell me? Perhaps I can be of some help?'

Tears were rolling down his cheeks uncontrollably. I took his hand in mine and guided him unresisting to my office.

'She is of the Luo tribe.'

The phrase fell like a stone in a still pond and even I was overwhelmed by its ripples.

A Luo girl and a Kikuyu boy!

The Luo tribe is spread around the warm shores of the great Lake Victoria, in western Kenya; the Kikuyu, as one knows, originate from the cold forest areas all around Mount Kenya's snowy slopes. Worlds apart, they have always been traditional enemies. Their people are physically and ethnically different. Language, appearance, customs, taste, habits, beliefs are totally opposite, deep-rooted. Even in this modern Kenya this was a very difficult – practically an impossible – match. Its implications dumbfounded me for a moment. I brushed them off.

'It does not matter, surely? She lives here in the town, far from the lake. Her people, you say, are reasonably comfortable. She is educated. She is emancipated. What does her tribe matter if she loves you too?'

'It is my mother. She will never agree. She has told me that she will curse me forever if I marry her. I want to marry her,' he added with a sob.

I looked at him, bewildered.

'If I do not have my mother's blessing as tradition asks, our children will be cursed forever.'

He looked at me through his tears.

The words rose from the Middle Ages, in one stroke wiping out years of education, Cambridge, books, culture, gold watch, computers, driving licence:

'She is uncircumcised.'

And so it came out, in my tame Nairobi office filled with photographs and books, the pine panelled walls that smell of the forests on the Great Rift where the wood came from. The place high on the Kinangop where Julius's mother lives, still now.

With her memories, and perhaps with her remorse.

He found her one day, that woman who was his mother, that peasant girl who had grown old too quickly through her labour, creeping about in the Nairobi flat that he shared with Olinda.

She shrank at his sight, hissing like a lizard, her headscarf half undone, trying to hide something in her apron.

He forced open her clamped hand, and recoiled in horror.

It looked like a small, dead animal, hairy and still. A thick, repulsive tangle of curly human hair, fashioned as a doll. Two rusty pins stuck out of its tiny head, like grotesque horns.

'How did you get Olinda's hair? How did you find us? How could you come in? You have never been here before.'

He pushed her away in fury, confused, afraid, angry, yet hopelessly dependent on his mother's fierce attachment, her resilient strength, her earthy power and the magnetic eyes which had subjugated him since his childhood.

I listened without ever interrupting him. He looked up through the veil of tears.

'The problem is that I love my mother. Love is not a feeling that she understands. She was married at puberty, soon after her circumcision, in exchange for a few goats and some gourds of beer, to a much older man she did not even know. He drank. He used her and beat her up every night. All day she worked the *shamba* cutting the red soil with her *jembe* to plant beans, potatoes and maize, with me and then my brothers and sisters tied to her bent back.'

He went on.

'When the coffee was ripe after the season of the rains, she went to pick it at sunrise with the other women. They were each given a large metal *debe* and they had to fill it with the red berries. They sang while they worked and those songs were my earliest music. She was quick and still so young, so supple, almost a child, she could fill even ten *debes* sometimes. They were paid a few coins each time. In those days the coins were made of brass and glinted like gold in the sun; many still had a hole in the middle below the potrait of King George, so the people could put a leather string through them, and hang them round their neck. No one had purses, nor pockets in those times.

'In the afternoon she chopped firewood from the forest and tied it in immense bundles, which she carried on her back: she had to strap the new baby to her chest instead, so it could nurse. She smelled of milk and sweat and wood smoke, and we felt secure in her warmth. The load was held by a leather strip tied around her forehead, and it was so heavy that it left a permanent ridge-mark over her brows. Loaded like a donkey she staggered home after dusk to cook our meal.

'We lived, cooked, ate, slept in one smoky room without a chimney. The fire's smoke choked us and gave us red eyes and sore throats; we all caught pneumonia sooner or later as the breath of the night outside was chilly, when we had to go out and relieve ourselves in the pit latrine in the bamboo thicket. She gave us a brew of honey and some bitter herbs and we were cured. But it was cosy, safe and warm inside our mud hut, like a den in the forest. There were scorpions and centipedes in the roof, sometimes black house-snakes, but it was too smoky for mosquitoes and too cold for malaria and too dark for flies. It was fine. There were only two makeshift beds. We children all shared a mattress filled with straw on a low pallet made of rough branches.'

I listened without interrupting him, fascinated.

'My father drank *changaa* until he recognized nobody. The whites of his eyes became yellow and stared unseeing. I tried not to hear their noises in the dark. But it was never really dark with those red fizzing embers. Once, when he beat her up more wildly I stood up to him. I was six years old. He knocked me off towards the wall, blinded by fury. For a few days I could not move, all my body ached. Soon after this, my father left us to marry a younger girl.'

He went on.

'My mother never accepted his second wife. There were terrible scenes and she screamed; she never cried. Tears were not something traditional for the Kikuyu girls in the old days. She was intelligent, although she could neither read nor write. As a child she had learned to sing hymns at the Mission church, she had a clear high voice and could remember sounds. She never knew what she was singing though, apart from the Kikuyu hymns. She never learned any English. I went with her. That's how I learned to sing.

'She cleaned the church for a few shillings a week. She began to spend much of that money and the coffee cents with the village *muganga*. She did spells against the other wife. She could not see the contradiction between the Christian god she served and the pagan spirits she summoned to her help.

'The *muganga* was a very old man covered in monkey skins, with long hair in matted locks. He wore bizarre ornaments of teeth, dried hoofs and bones, and he stank of coagulated blood and strange herbs. Purses hung from a leather belt, he wore sandals made of raw hides, he scowled at us with red wild eyes, and I was afraid of him. He lived alone on a hill out of the village with a few black goats, next to a large *mugumu* tree. All children were terrified of him and we ran away when he approached.

'To perform her sorceries, my mother had to procure certain items that he ordered her to find from time to time. To perpetrate a really powerful incantation a goat was always slaughtered, her bowels examined and scattered with strange rituals, potions of roots and powders boiled up and, sometimes, for a sickness or a death spell, weird dolls made to resemble the person who had to suffer. You know, people died of spells very easily. Or sometimes they went mad. They still do. Those dolls had to contain a bit of hair of the victim, or perhaps some fragment of his fingernails, a piece of material from his clothes. The puppet had to be placed secretly somewhere in the victim's house or on her or his person; a pin had to go through the part of the body where the disease was meant to appear.

'Occasionally one saw someone moving around in a trance, as if sleepwalking, with a vacant stare, and we knew that a spell had been put on them. If the *muganga* who was responsible was not found and compelled to give an antidote to neutralize the witchcraft, the person lost all interest in life, stopped eating, and went off like a candle.

'My father died a few years after. They found his mangled body in a ditch one morning. The hyena had found him first. They said he had fallen in drunk. Perhaps this was true. But he had been seen looking wild and going around in circles, and I wondered.'

'Your mother would not do that to your girlfriend. Surely not. That's evil, Julius!'

He looked up at me without answering.

Then he murmured something in a half voice, something quite unbelievable, which I was not sure I had heard correctly.

He asked me then if he could be in Laikipia for two weeks, staying in the empty house overlooking the forest at Enghelesha, far from the crowds and problems, unreachable even by his girlfriend.

He had to think, rest his soul, reflect, be alone.

I saw him a few times. He was aloof, remote more than ever,

closed in a world of his own. He had lost much weight. As if he no longer cared for his appearance, he had begun to grow a beard.

I invited him for a picnic at a dam one Sunday, and he came. He did not eat. He sat on a rock at a distance, throwing in the conversation.

'What are you doing all day?'

'I read.'

'What do you read?'

'I read Shakespeare.' And with a shadow of his old humour he added: 'It is so grand reading Shakespeare aloud looking down at the Rift Valley at sunset.'

I could not agree more.

Shakespeare, Cambridge, the *muganga*, the rural superstition, the deeply rooted beliefs, his English tutor's influence, his studies, his Luo girlfriend from the great lake, this new overwhelming love and passion that he was so unprepared to handle, his terrible mother.

I was seriously worried for Julius.

I never saw him again. It was three months after that Sunday when my phone had rung with the tragic news.

'It's Julius. He has committed suicide.'

Sitting now on my white sofa, gulping down whisky absently like one who never drinks strong alcohol and could not care less about the consequences, Anthony went on talking.

'I had seen him a few days before. I found him in his room. He sat on his bed and looked at me with empty eyes, drawn, not listening. He had grown an untidy, long beard. He looked like someone under a witchcraft spell.'

I bolted.

'He had a book of Shakespeare next to him. He was reading *Romeo and Juliet*.'

A pair of star-crossed lovers.

A pause.

'He swallowed rat poison. His mother was furious when he died.'

'Furious? Not sad? Not desperate?'

'Angry, as if something had gone wrong.'

Wrong?

What was it that Julius had said to me that day long ago, in my office? It was important that I should remember. Something about the spell. The hair. Olinda's hair. No. Not Olinda's. The spell was not for her. It was for Julius. So that he would lose interest in her, leave her. The spell went wrong. Everything now was clear.

He had lost, instead, all interest in life.

His voice came back from distances to whisper:

'It was MY hair in the doll.'

The Rain-Stick

For Isabella

There was a roaring in the wind all night;
The rain came heavily and fell in floods.
                    WILLIAM WORDSWORTH,
                    'Resolution and Independence'

The aeroplane came to a halt in a cloud of red dust, roaring its engines while it turned towards us. The girl jumped out, dressed in beige linen, beautiful, still pale from the journey and the strain of an American winter.

'I heard you had a drought,' were her first words, and she held out to me a long object, wrapped in brown paper.

She looked around. On every side of the strip at Kuti, skeletal shrubs and dusty yellow stubs of grass painfully broke the hard crust of murram, mute witnesses to her words. It was the second year of a harsh and painful drought. Thorns and dust were all that was left of the green, vibrant bush.

'Here is my present; it is a rain-stick. I bought it at an American Indian shop in New York. They say it is infallible.' She grinned. Her stunning eyes glinted with gold mischief.

'I hope it works. It looks as if you really need it.'

It was the morning of 24 December, and in my large verandah, below the thatched *makuti* roof, the Christmas tree glittered its unexpected tinsel reflections in a merciless equatorial sun. The air was hot, and still, and dry, with no hope of moisture. The sprinklers turned and turned slowly, tired on my lawn, spraying the flower beds with perfunctory jets of water which were instantly absorbed by the thirsty soil. Birds flew in to shower, shaking and ruffling up their feathers with chirping trills of pleasure; and the go-away birds cried from the treetops their raucous noon calls, querulous as ever.

I did not want to unwrap the rain-stick yet, as it was a Christmas gift, but when I took it in my hands an extraordinary noise of running water came from it, a liquid sound of droplets streaming through its length with a passionate intensity. It sounded like a shower of rain falling on a thatched roof with powerful abundance: a forgotten sound. If anything could attract rain, it was surely this.

When had our last rainfall been? It did seem ages. The worst drought ever was killing Kenya's crops, animals and people. A famine of unrecorded cruelty was ravaging the Northern Frontier, where parched camels' carcasses lay on the sand beside desiccated water-holes in the quivering heat, and people died like ants day after day, of malnutrition and thirst and nameless disease and lost hopes.

It had been a terrible time. Over a period of two years with practically no rain our water-holes had dried out. Islands had emerged from the large dams that had reached their lowest level, and were rimmed with dusty patches of unhealthy green reeds where the water had once been. Algae grew on the surface suffocating the life below. Clouds of dead tilapia apppeared bloated on the surface, poisoning the depths. Our cattle had lost condition and many had to be sold; the remaining animals struggled to survive on a sparse diet of sticks and dust and salt. Buffalo were found dead every day. Skinny gazelles, with sad eyes and unhealthy coats, stood in forlorn groups, licking the dust. Even the mighty elephants looked thin, their ribs showing beneath the corrugated skin, and each night herd after herd persistently attacked my garden, the only oasis of green in a vast area.

The ranch was dying, and there was nothing we could do. It was late December. No rain could be expected before the end of

April. But there was no way we could survive until then. We needed a miracle; I could surely do with a rain-stick.

The rain-stick, when I finally gave in to my curiosity, was a simple section of bamboo, thick and decorated with a fringe of red and black silk ribbon. It had been crafted cleverly, with much thought. When I touched it and turned it upside-down an invisible series of seeds fell through it and touched the hidden thorns which perforated its length at intervals, interrupting their fall and resounding like secret castanets; it was an uncanny sound of rain on a roof-top. I brought it in triumph to the kitchen, and explained to the staff its infallible spell. '*Ni miti ya mvua. Natoka ngambo, mbali, kutoka mganga ya asamani. Ni kali sana.*' 'It is a stick to make rain. It comes from far away across the sea, built by old witch doctors. It is very powerful.' With the unquestioning African faith in talismans, they instantly believed me. '*Tasaidia sis,*' Simon the cook declared seriously, accepting its magic with simplicity. '*Asante sana.*' 'It will help us. Thank you.'

They all nodded wisely and touched, in awe, the occult instrument of rain.

I shook it up in the air with naive emphasis, begging the gods for rain. Rachel and Julius clapped but did not laugh. One does not play with mysteries.

That night, believe it or not, the first tentative drops began falling without much enthusiasm on the tin roof of the bedrooms. January is one of the driest months of the year in Kenya. It was highly unlikely that any rain would fall for months. I woke up at the noise, lit a candle and looked out of my window. A cool wind blew, and with it came a distracted spray of sparse warm drops. I took the rain-stick from my bedside table and shook it a bit more for good measure.

Was it an impression? Did the sound of rain gain momentum? A coincidence? A delusion? It seemed so, for the next day the sun was as blazing hot as ever. But a faint scattering of raindrop marks on the dust of the driveway was proof enough for my staff, who came to ask me if I could please beg again with the rain-stick.

And so it happened. It began surreptitiously: a difference in the wind; masses of dark grey clouds coming from the east, and passing through without stopping, like flocks of foreign sheep in the silent sky. A sound of far-off thunder in the early mornings, and in the evenings a sickly rim of silver on the horizon, blackening the sun. A stillness in the air, a chill suddenly creeping into the hot noons with drifting shadows.

Then the news came that it was raining up north, over the Chalbi desert. On a ranch called Borana several inches fell in one day. People murmured in wonder.

We had arranged to take a few days off, and go to a camp in Samburu, called Kitich, on the Mathews range. The Mathews were dry mountains with forests and ancient cycads, and rivers of great beauty. I flew in with Aidan and Sveva, while Jeremy Block followed with his plane, his father and other friends. The mountains appeared shrouded in clouds, and the closer we approached the more we could smell rain. Even from the air we could see wet patches on the tracks.

When we landed, the strip was so wet that, had it not been sandy, we could never have made it. Jeremy took off again immediately for Colchecchio ranch, which belonged to my old friend Carletto, to ferry in some more friends from there to join the party.

'You'll never make it back, with all this rain!' we joked.

Rain in Colchecchio in January was out of the question. We all laughed, but somehow we all felt a premonition, and as I looked up, the clouds seemed to gather faster, covering in mist the mountain tops. We drove in two packed Land Rovers, slipping through mud, often sliding sideways, and with great difficulty we twice crossed a river which seemed to get fuller by the moment. Our guide shook his head.

'If it goes on raining tonight, the track will be blocked, and we shall never be able to drive back through the river.' He looked confused. 'We have never seen anything like it since I have been here.'

Along the road we met a thin black goat, happily herded by two craggy Samburu elders dressed in red. Red and white glass bead ornaments glinted from their earlobes, like tiny antlers. They wanted a lift, but there was no way two extra people and a goat could be squeezed in with us. The Samburu went on their way, waving goodbyes.

'It is the goat for the leopard,' explained the driver.

We observed it with curiosity. It was really skinny, with a bloated stomach.

'They will slaughter it, and then we shall hang half of it from a tree.' He smiled broadly: 'We shall eat the other half.'

'The leopard is very hungry. He shall surely come tonight to eat it.'

And we would watch the leopard coming to the bait.

We looked at the innocent goat trotting along unknowingly to her fate. Strangely, somehow, there was no real cruelty in the scheme, as the leopard would take a goat each night anyway, and the camp would buy one for the staff stew, and the fact that this one was designated for a sacrifice seemed to fit well with the Samburu's business. They all needed cash after that long nasty drought.

We proceeded so slowly in the mud that by the time we reached the camp we found the two old Samburu there, leaning on their sticks. The unfortunate goat had mercifully been already *chinjad*. Our tents were new and comfortable, near the river bank. The river was full, and to our surprise huge tree trunks were being carried off on its tumultuous waters, proving that unusually heavy rains must have fallen on the forested mountain tops.

It started drizzling soon after we arrived, and by the afternoon it was raining so hard that we could not leave our tents. Some even had to be moved, as the ever-rising water lapped at the flies of their verandahs. The only dry places were our beds where we repaired for most of the time, and I managed to read again, after years, the entire unlikely story of *The Portrait of Dorian Gray*.

Only Aidan, unworried by the elements as he could be, took off with his rucksack in the rain and came back in the dark, soaking wet, his *kikapu* filled with strange succulent plants.

We crept out in the evening for dinner, giggling at the absurdity of this adventure, and during the first course in the dripping mess tent, our attention was drawn to the lit tree across the river, where half the goat hung from a branch. There was a large genet cat on it, pulling off chunks of meat with hungry determination. We stood watching until suddenly from the dark foliage a shadow jumped up into the tree, the branch swayed, and a leopard crouched in all its mystery in the light, grabbing the carcass with its claws.

The genet cat's mistake was its greed. The leopard, impatient of competition, swerved to one side, and with a casual jerk snapped its fangs on the genet's neck. It trapped the genet in its powerful jaws and shook it twice; we saw a shiver pass through the grey lithe creature, and its limp body was thrown carelessly off the branch, back into the darkness, leaving us breathless. As if nothing had happened, the leopard went back to the goat, to feed on it, indifferent to the heavy rain, until only a few bones were left.

This was the major event of the wettest two days of my life. Jeremy was never able to fly back to join us, because Colchecchio – we heard on the camp's radio that first evening – was under water, with thick fogs, and a constant deluge.

Eventually, we had to leave on foot in the torrential rain, wading through the river with water up to our necks, the luggage balanced perilously on our porters' heads. We walked all the miles back to the aeroplane, which stood on its sandy strip on the hill, glistening with moisture.

We flew back through steamy mists over new swamps and

muddy, swirling rivers which filled the *luggas* and inundated the sands around for miles. Five inches of rain had fallen over a day and two nights. Two of Carletto's dam walls had collapsed at Colchecchio, and torrents of water flooded the plains where zebra and giraffe ran bewildered.

The rains had come to Laikipia too. Night after night I lay on my bed listening to the water gushing down the roof. Our big dam filled in a few days. As if by miracle, grass sprouted again on the barren savannah; slowly at first, animals regained condition, and the elephants for a time forgot my garden.

We found that rainstorms had spread through the country, causing unprecedented floods everywhere. Bridges had collapsed, trains had gone off the rails, rivers and lakes overflowed, and panels of experts were puzzled to find answers to this extraordinary change in the weather pattern, unrecorded in human memory. Fantastic new explanations appeared in the newspapers every day. Even the BBC World Service, the Bible of all good Kenyans, spoke about it.

January 1993 was declared the wettest January in the history of Kenya. The final explanation was that a cyclone bound for Madagascar had, for no apparent reason, changed route and poured its avalanches of water onto the parched soil of Kenya. Wherever one went, people exclaimed in surprise.

But we, at Ol Ari Nyiro, knew better. On my bedside table the rain-stick lay enigmatic, and many of the ranch people came over to thank it. With knowing glances, wherever I went on the ranch, they alluded to its powerful medicine. Even the Pokot came to know about it, and they suggested that every season I should go with it where rain seemed to be most needed.

Now it is March, and the dams are again beginning to dry out. People start whispering that I should use the rain-stick once more. But I prefer to wait, as the gods should not be bothered too often. I shall resist and wait to the end of April, when the rains are meant to come anyway. And when the sky begins to darken with clouds from the east, I shall again, with reverence, use the rain-stick from the Red Indians to attract rain to this land once more.

The rain-stick belongs to myths and legends; and as one has learnt in all the tales one has read as a child, magic powers should not be abused, lest one should lose them.

Part V

A TASTE FOR ADVENTURE

Why an Elephant?

O my America! my new-found-land.

JOHN DONNE, *Elegies, On Going to Bed*

1

'*Dear Mrs Gallmann*' – the letter said – '*Thank you for allowing us to stay in the Education Centre in memory of your late son. I wish I had met him. I loved everything, but mostly the wildlife. The first animal I saw was an elephant. I had never seen one before. It was very big, I could see him from far.*

Anne Wanjoi, Nyahururu Secondary School, Ngarua'

On the large rock set in the middle of the yard, a red hibiscus beckons in the breeze.

'*Felix qui potuit rerum cognoscere causas*'*

says the inscription from Virgil, on the brass plaque. And, underneath:

IN MEMORY OF EMANUELE.
VENEZIA 1966 – LAIKIPIA 1983.

Two acacia trees, one small, one large, the symbol of survival and of hope, are engraved below.

* 'Happy he who can understand the causes of things' – Virgil.

223

A short life, an entire story in a few words.

I am at the Laikipia Wilderness Education Centre, built of stones and thatch in the heart of the ranch where I live and where my son died and was buried at seventeen, one sunny day of April in 1983.

Yet, in the agony of his loss, slowly, like the secret buried seed which finds the strength to sprout, forcing its way up towards light and life, I found the key to survival. I understood that Emanuele, like Paolo before him and all the ones who had gone ahead faster than us, had simply changed dimension, and that, if I could no longer see his body, I could always feel the power of his essence, of which I was myself part. There were no boundaries.

I founded the mission to work positively for as long as I could, to justify my existence and my survival by using my opportunities to make a change.

I soon discovered that there was no end to the possibilities brought about by this new energy. Emanuele had gone to join Paolo, but the power of his love and memory, reflected in the African nature that he had loved, became my inspiration to create something long-lasting, that time could not tarnish.

Walking alone on the hills and in the valleys of this place he had loved above all, I decided to open its door to other young people like him: enthusiastic, fond of the wild, with a sense of purpose, so that they might learn to understand, and so to respect, Africa.

Only two generations ago Africans knew all there was to know about their environment. They lived in harmony with their surroundings, which in turn provided them with food and medicine. They survived by hunting and fishing, collecting plants and tubers, gathering honey and wild berries, without destroying the subtle, fragile balance of their ecosystem on which their very existence depended. In those days they were familiar with their roots, they had pride in their customs and traditions, they valued their world, and they knew elephants. Those days are no longer.

Unless the new generation, now converging on urban areas to be educated in alien skills imported from a far-away world, is given the chance to experience the wilderness which was their immediate ancestors' familiar domain, to be proud of their land and its unrepeatable beauty, there will be no hope for the ecology of this continent. The forests will go, and the plants and the bio-diversity and then the birds and the animals, and Africa will remain a mangled landscape, ravaged by droughts and floods.

I believe that childhood impressions live in us to guide our adult behaviour. This is the principle that guided me when I created the Wilderness Centre.

On a day like any other, sitting on the look-out built at the end of the promontory on which the school stands, on the shores of a small dam covered with clumps of papyrus and frequented by water birds, I watch a herd of impala coming gracefully down to the water to drink. Egyptian geese fly off, screaming, in a flutter of wings. Behind them, a water buck, and from the *lelechwa* shrubs, one after the other, file the elephants.

Standing on the shore, in their colourful school uniforms, a group of African children watch in awe.

2

The young man in the neat, well-cut blue suit held a card with my name, misspelled.

'It's me you are looking for.' I introduced myself, and he flashed back a frank open smile which split his pleasant dark face in a happy grin.

'I was not quite sure of the spelling. I hope you had a good flight.'

I was tired, coming in from a few days of work and promotion in Seattle, still jet-lagged, but looking forward to exploring this part of California. The air was warm but surprisingly dry. The sun shone. From the sky Los Angeles's suburbia had looked like an arid, overcrowded third world city of endless, low, identical buildings.

I followed him through the crowd.

'Let's take my luggage and then we go.'

He took my computer case and glanced quickly at me.

'I cannot place your accent.' He was self-assured. 'Where are you from?'

I laughed.

'I am originally Italian. I thought everyone could detect my accent as soon as I open my mouth.' My Italian accent has not faded after more than twenty years in Africa. I am rather proud of it, and in any case there is not much I can do to improve it.

After all, I was once Italian.

'Oh I love Italy. I flew there with Alitalia and they said "*Signore e Signori, benvenuti a bordo*".' We laughed together. 'It sounded so nice I would have liked them to repeat it again and again. Where do you live in Italy?'

'Well, I actually no longer live there. I live in Africa.'

He stopped abruptly.

'Where, in Africa?'

For the first time I noted that his accent was not American; I had taken for granted from the beginning he was an African American. Now I looked at him better.

Slim, of medium height, intelligent eyes, small ears, regular, slightly familiar features.

'I live in Kenya.'

Even before I said it I anticipated what was going to happen, but still his joy took me by surprise.

'*Oh jambo! jambo jambo mama. Hata mimi natoka Kenya. Mimi ni Jaluo. Halla bahati mzuri sana kukutana na wewe hapa.*'

'I come from Kenya too. I am a Luo tribesman. What strange good luck to meet you here.'

How lucky indeed. That the person sent to meet me in Los Angeles where I went because of the planned feature film based on my autobiography, *I Dreamed of Africa*, is an African and a Kenyan, and this by pure coincidence, cannot be a coincidence. It is an omen. A fantastically good omen.

That set the mood.

I am delighted; a chance to speak Swahili during this trip.

We find the luggage, he goes and gets the car. A comfortable, dark blue, normal-looking limousine, not, thank goodness, one of those over-long vulgar affairs.

We settled in and I asked him about his life. How and why did he get here?

He drove with easy assurance.

'I had a dream,' he said, 'of becoming a business administrator. I wanted to go to college in California. That's what I am doing now. I drive to help pay the fees.'

'Your family must be quite well off to send you over here. How did you come? You must have connections.'

He grinned at me in the rear mirror. 'Not a penny. Not a single connection, apart from an older friend who was working here. I had twenty cents in my pocket when I arrived.'

I considered his new suit, a good tie, clean, perfectly ironed shirt. A certain opulence transpired from him.

'How long ago was this?'

'Three years ago. I slept on the floor in my friend's rented room for months. And, to survive, I cleaned toilets. A job no one wanted. I cleaned toilets for cash. I was not allowed to work, to start with.'

I was transfixed.

'And then?'

'And then, after months of cleaning loos, I started cleaning windows. I saved every penny.'

He looked at me in the mirror.

'From windows, I graduated to selling things in the street. Odd jobs. Whatever came up. It was not easy, but everything is possible in America.' He paused, there was pride in his voice.

'Now I have a green card. I go to college. I graduate next year.'

We were crossing many roads and finally went down the freeway towards Santa Monica. The ocean glittered, blue and turquoise, and on the white beach people jogged.

'It must have been hard to start with,' I murmured.

'Not really. I knew what I wanted to do and there is no work too humble, when you do it for a purpose.'

I loved that attitude.

'I work for an insurance company too. I study mostly at night and at weekends. Life is great.'

'Your family must be very proud of you back home.'

'They are. My brothers and sisters. Back in Kisumu.'

What a step from the busy, provincial little town on the shores of Lake Victoria to this chaotic modern symbol of America's metropolis.

'Do you miss Kenya?'

'I DO in a way. Yes. But I will go back next November. I saved enough for the ticket.'

I was more and more amazed, proud of him. He achieved the practically impossible.

We had reached Santa Monica. The ocean sparkled, small waves crowned in foam, and on the beach, Walkman and headphones on, a young couple ran, holding hands. A few youths dashed fast amongst them rollerskating.

'Joseph' – he reminds me of another Joseph, the same joy in his eyes – 'did you ever see an elephant back home in Kenya?'

I had managed to surprise him. For a moment his self-confidence was out of balance.

He braked, almost swerved.

'An elephant? Why?' His eyes were round with surprise. He shrugs.

And here, in his reaction, in his bewildered answer, was the drama of Africa's environment, the loss of identity of the African people, the reason why the environment is not safe, and my major reason for being in America.

To create awareness. To ask for help.

Local people in Africa cannot afford even to see their own wildlife. As if in Italy we could not afford to enter our churches, to be inspired by our monuments, to visit our museums, to contemplate the Roman ruins, to experience the art which makes Italy great. And this is why I have built the Wilderness Centre.

'An elephant? Of course not. How could I possibly see an elephant? How could I afford to see one?'

I have invited Joseph to come and stay in Laikipia when he returns to Kenya in the fall, to see the sun rise on the savannah in a deafening concert of birds, to walk to the waterfalls on the edge of the Great Rift Valley amongst palms and fig trees, and to stalk his first free elephant, king of the bush, so that when he returns to Los Angeles, the memory of the pachyderm which is part of his heritage will make him walk tall.

I know well that like thousands of other youths who have visited Ol ari Nyiro, he will never be the same, because he will have seen an elephant moving majestic and free through the African bush.

Birthday in Turkana

My birthday began with the water-
Birds and the birds of the winged trees flying my name.
DYLAN THOMAS, 'Poem in October'

When Sveva was about to be ten years old, I asked her:
'Now, tell me. Where would you like to celebrate your tenth birthday? It is a most important one, the first one with two figures.'

She looked at me with those turquoise blue eyes, like Venetian trade beads, which had been Paolo's.

'Choose something special,' I continued, 'a symbol of what you would like to do for the next ten years. And tell me why.'

I did not know what she would ask. But as I knew she had come from a wild seed of unpredictability, I anticipated that it would not be obvious – not a party with cakes and music, pony rides and fancy dresses, confetti, presents for each little friend and perhaps a treasure-hunt.

There was the time – she was about eight years old – soon after our Turkana companion and guide Mirimuk had died, when she asked to spend the night of a half-term out in the bush with me in Laikipia, without a tent. We went alone to Luoniek dam, and, after eating and telling each other stories around a fire of *lelechwa* roots, we had slept on a large mattress on the ground, under a mosquito net tied to a small tree as an illusion of protection.

In the middle of the night a noise woke me up. Grey and

230

massive in the light of the moon, hovering above us so close that he could easily have touched us, was a large male elephant. He had stopped next to us to pee, and was attracted by the net flapping in the breeze. Totally still, head tilted on one side, he seemed to be listening. His tusks stood out, white in the moonlight.

I caught my breath and looked at Sveva. She was fast asleep next to me, curled up in her blanket, her fair hair spread on the pillow, and she appeared innocent and vulnerable, so terribly small below the tall elephant shadow. I moved my head close to hers and murmured in her ear, while squeezing her arm gently:

'Wake up. There is an elephant here. Do not move. Be ready to run into the car when I tell you.'

My Toyota, with the back door open, was parked nearby.

Unlike children normally shaken from deep sleep, she opened her eyes instantly and fixed them into mine, slightly glazed for only a moment. Then, without moving her head at all, she turned them up and considered the elephant. She was not worried by what she saw.

'We'll be fine,' she murmured with a smile; then she closed her eyes and in no time she was asleep again. Soon the elephant moved away quietly. I shall never forget that incident.

Now she considered my question. It did not take her long. She smiled brightly, glowing as ever.

'I would like to go to an island in Lake Turkana, because you said it's wild and beautiful, and Papa Paolo and Emanuele loved it. It will be right to be in a place where I have never been yet, because there will be nobody there, apart from us, and because for the next ten years of my life I want to see special wild places where few people go.'

'That's a deal,' I said happily. It had been ages since I had travelled to Turkana myself.

Turkana of the savage, torrid winds, like the breath of hidden giants; of the basalt rocks and immense waters, and crocodiles and huge fish; of solitude and silence, and spectacular landscapes:

I longed to go back. But, to an island? How? There were no islands large enough to have an airstrip, and few real airstrips round the lake apart from one at the oasis of Loyangalani, and another in the north near Ethiopia, at Koobi Fora where Richard Leakey had his anthropological dig. And we had no boat. There was only one island flat enough to land on, and was the remote, deserted South Island.

So strong were the winds and so barren the waterless dunes, that no one lived there. Only untamed Turkana fishermen, naked or clad in the flimsiest loincloth, reached it on fishing *safaris*, after rowing through turbulent, crocodile-infested waters, in their wooden canoes packed with harpoons and hand-made sisal nets.

There was no strip, just an almost flat area, like a curving track through the rocks and rare thorn trees, with a bend and a dip in its middle, as if drawn in the black sand by the swishing tail of a dinosaur.

I had heard about the island's extraordinary beauty, its dramatic landscapes, the difficulty of reaching it, the danger of landing there in high winds. For years I had heard about its bewitching spell.

Among the places I had always wanted to see, but had never found the right person to go with, or the right circumstances, South Island in Turkana was certainly first. But among my friends with an aeroplane, there was not one of whom I would dare to ask such a favour.

So, South Island had remained beyond my reach. As with all places long coveted, the yearning made it more precious, unattainable. As with all places of essential simplicity and stark spiritual quality, I felt the adventure of its discovery should only be shared with someone I felt deeply in tune with.

But now there was one person in my life who could help to make this mad dream come true. Aidan.

He was the one who walked alone, exploring virgin mountains and unvisited deserts. He loved travelling through new country with only his camels and camel handlers for company. He knew the African bush, and he knew the sky. He flew his small plane with legendary ease and could land anywhere – on a road, on a beach, on the sand, by the light of the moon or of a hurricane lamp.

Our relationship had known the test of time and the agony of separation, but when it pleased the gods he had flown back into my life. I was allowed to see his eyes by daylight, and to walk at his side in the sun: together we could smell dry grasses, dusts and growing things. Now was the time of travelling together to strange places.

There came a week of walking with his camels through deserts, and the bewilderment of the desert kiss. There was the time he flew me for hours over parched hills to a place of antique Muslim dignity and traditional hospitality, a small town on the border of Ethiopia.

There was a day when we landed on a forsaken strip in the middle of nowhere, and walked along a dry river bed to find a biblical well, herds of camels and goats, and wild herders in turbans and loose *shukas*. And there was the night when the noise of a light aircraft landing on my strip at Kuti had woken me up, and I had run out, followed by my dogs, to find him walking up my drive.

'I thought there was just enough moon to ask you for a moonlight walk.' He had smiled at me. Sveva, who had been spending that night in a tent in the garden with a friend for fun and adventure, told me next day they had believed a rocket from another star had landed on our strip.

Aidan was, beyond doubt, the ideal companion for South Island in Turkana. Sveva's tenth birthday was the ideal occasion. My daughter and my man: what a perfect combination.

The aeroplane skidded in the black sand towards a bare hill tinged with orange, and with a last roar the engine stopped. Then there was only the silence of the island, coated with the fire of the evening.

I opened the door and jumped out. The warm wind of Turkana touched me again, after many years, bringing the soft soda smell, the cry of a crow and a wave of memories. The wind caught Sveva's hair in a sunny swirl.

'Thank you, Mamma!'

I turned round, shaking my fringe from my eyes, and as far as I could see there was only water, volcanic sands, black gravel, yellow hills and mountains devoid of any human life. We were the only people on earth, and the first ones, and the last ones.

I squeezed Sveva tight; her head already reached my chin; she would be very tall one day soon.

'Happy birthday, *amore.*'

So, this was it, South Island in Turkana. I may live to be a hundred or I may die any day. To my last moment I will carry with me the memory of those enchanted days and nights in Turkana with Aidan and Sveva. There was talk of war in the world during those days, tension growing in the Middle East, Iraqi threats, the invasion of Kuwait. The world was waiting ready to start a bloodshed born of mad pride and greed. It did not matter in Turkana.

We discovered a large twisted acacia on the side of the hill, growing alone and wise, barely out of the reach of the winds. And in turns we carried down to it what we had brought: mattresses and mats, jerrycans of drinking water, a cool-box full of food and a basket with the birthday cake – a chocolate heart that Sveva's *ayah*, Wanjiru, had packed lovingly, and to which Simon had added ten blue candles, and a handful of pink bougainvillaeas, for *maridadi*, for beauty.

We set our belongings below the thorn tree, closely observed all the time by a couple of fantail ravens, whose territory we were clearly invading. Yet they did not seem to mind, and in fact eyed our food with eager fixity and undisguised anticipation. It was clear that they would become our constant companions, and to captivate them from the beginning, and show them we did not mind, I threw them some bread. Unwanted ravens and crows, I knew from experience, could become persistent pests, creeping about and spying for any chance to land and steal morsels of unguarded food.

We went for a swim in the warm soda water, down the sands of grey lava which held the sun's warmth long after it had set, walking into the lake awkwardly on slimy stones, always looking out for crocodiles. From the top of a rock we watched a large Nile perch darting around in her underwater territory. We rested on the hot sand and found crystals hidden in it like lost jewels.

At night we lit a fire of twigs which sparkled in distracted bursts with the wind blasts. We sat, with our back to a fallen branch of the acacia that was large enough to create a special protected corner for the three of us, cosy with pillows and straw mats. We opened a bottle of champagne and ate cold pasta; and, while a leg of lamb skewered with rosemary and wrapped in foil roasted on the fire, we told each other stories of the past.

Finally, with difficulty, we managed to light the ten large blue candles. The wax fizzed, melting quickly in the great heat. Sveva blew them out faster than the wind, and we all kissed each other, while our friends the ravens sang their raucous approval, guarding with tilted heads our celebration from the tallest branch.

No other little girl, we knew, was spending her birthday in such an improbable way, with only the wind and a lake for company. And the gods were smiling.

We spent two nights under that old acacia and shared the same blanket. Aidan climbed the parched hills looking for rare plants. We found bleached fish bones and brought them back to Simon, whose ancestors once used to live in Turkana.

Someone recently told me that they went to South Island and found some blue wax stuck to a branch of a solitary acacia tree. They could not work out what it was.

One Day, in Kiwayu

A man who is not afraid of the sea will soon be drowned,
he said, for he will be going out on a day he shouldn't.
But we do be afraid of the sea, and we do only be
drowned now and again.

J. M. SYNGE, *The Aran Islands*

The heat of the sand penetrated my tired body and I dug my toes in it, stretching my muscles, content, finally relaxed, conscious of having the whole earth as my bed. The sun warmed my back. Through half-closed lids I could see the small waves of low tide lapping the shore. The whiteness of the beach reflected a million hidden crystals, minute shells, and across the bay the island of Kiwayu was green, its lacy acacia trees filigreed with hanging liana. The rocks encrusted with oysters exposed by the tide were now rimmed with foam, and a wind had begun to rise. Clouds were gathering on the horizon: perhaps it would rain in the evening.

Today was my birthday, June the first, in Kenya Madaraka Day, a national holiday. We had flown to this idyllic stretch of sand not far from the Somali border on my return from a long overseas trip. Here I could rest before going back to my demanding work routine, get rid of the jet-lag, celebrate my birthday away from disturbances, thousands of kilometres far from civilization, where no telephone, no message could reach me. Even in Laikipia the problems accumulated during my absence would have been waiting for me. It had been Aidan's idea, thoughtful, caring as ever. Kiwayu was special to me.

A place full of memories of sand crabs dancing away on light legs, like my emotions, of small grey waves licking the shore, red suns melting on pearly waters, as in a floating mirror of coral and quicksilver. Of walking to the village of Mkokoni, where Paolo and I had once spent a blissful summer camping on the beach. Of Emanuele sailing out for cowries with the fishermen in the early morning, his thin arm lifted, waving goodbye. Of night picnics on the sand, sitting on large damp cushions of cotton *kangas*, smell of sea wind and long-gone youth.

I breathed the warm, balsamic sea-air deeply. Next to me, Sveva stood up.

'I have been asked to go out sailing. Can I?'

I looked at her long tanned legs, hips which had recently taken a new rounded shape, a body which seemed to have bloomed almost overnight, while I had been in America to launch my book, to talk for the Foundation, to work. She shook her heavy, honey-coloured mane of hair, one of her most striking features, and I realized with surprise that she was a child no longer. Gone was the chubby Mediterranean angel face, the fine silver gold baby-hair. Soon, any time now, she would be a woman. Soon she would be gone.

'Are you sure you know how to?'

With the sudden, resolute Paolo's look which I knew I could never fight, a determined glint in her oceanic eyes, she turned to me. A smile of full lips, Paolo's white teeth, the radiance of her glowing complexion.

'If I never try, I shall never learn. I'll be back for lunch.'

She ran off, a glimpse of youth and blonde and copper, to join her waiting friend. Only a short unease, a brief misgiving; and the tide of sleep and exhaustion, the giddiness of the jet-lag, the warmth and the rhythmic, hypnotizing sound of the ocean claimed me, and I was asleep.

I woke up slowly feeling stiff. My shoulders were burning and the sun was now over the zenith. I glanced at my watch and saw that I had been asleep for over two hours. I must get back to my thatched cottage, see how Aidan was, find Sveva. I was hungry now.

I stood and looked along the beach. A solitary figure was coming towards me, almost running.

The wind was stronger and the choppy sea looked green and murky, with large ugly waves heavy with seaweed.

I sheltered my eyes with one hand and saw that it was Aidan. His tall frame bent forward against the wind, he moved with a curious speed. There was a sort of urgency in his step which I could detect even at that distance. With a quickening in my heart I ran towards him. A greeting smile died on my lips when I saw his face.

He always came straight to the point. His voice was more serious than ever. Its gravity added to my apprehension.

'You'd better come quick. It's Sveva. There has been an accident. I am afraid she doesn't look too good.'

Instantly, my mouth dried out and my voice with it. A silent scream rose within me, from those depths of agony I had visited before, closing my throat, carrying, like a chill breeze, the familiar sense of horror, impotence and impending loss.

Not again.

Oh no. No. Not Sveva. Not Sveva, oh gods in heaven.

In moments of despair my soul goes begging to the spirit friends I have in the world of beyond. The ones I have loved and lost. The ones who would always care for my tears and for my pain.

Paolo, oh Paolo. Emanuele, Emanuele, please. Nonna, Nonna. Please. Not Sveva, please, not her.

I stood only an instant irresolute, a painful rigidity choking me. Aidan's blue eyes were slits of worry, his forehead closed in a frown. He held me up.

'. . . Not really serious. It's her eye. Her face. She was hit . . .'

I was running. Why was the burning sand so cold at high noon that day in Kiwayu under my flying feet? Why was the distance between me and the cottage in the raffia palm grove interminable? Why was the beating of my fear like a pulsating drum which deafened any other sound? And all along, in the back of my mind, the knowledge that it was now afternoon and a national holiday, that we were hours of flying from any medical help. We could never make it before dark.

Aidan could have flown in any weather, with any moon, but tonight there would be no moon at all. We could not fly in total darkness over lightless lands.

I prayed with no words to the ones who would care enough to make a difference.

Please please please Paolo, Emanuele, Nonna, per piacere aiutatemi.

The familiar, determined coolness which terror carries with it took hold of me again and I split in two. The other Kuki ran. The sun was dark and alien like a sun in the moon. *Run Kuki run.* The last rise over the last dune. A tangle of sea vines twisted around her ankle and almost tripped her over. The entrance of the room, sheltered by bright cotton *kanga*. The palm matting on the floor cool, soft under the soles of her feet. She held herself a moment against the door-frame, short of breath, and she was there. Back to the here and now, and to herself.

In the dark it took some moments for my sun-dazed eyes to adjust. Sveva lay on my bed, wrapped in a red wet *kanga*. The young man who had taken her out sailing was crouched at the bed's foot holding his head. Ignoring him, I ran to her, calling her name, and when she heard my voice she moaned.

She is alive. She is alive.

I looked.

Her wet hair was plastered to her skull. Below it, what had been her angel face was red and swollen, covered in blood. A crescent-shaped open cut scarred her left cheekbone, and blood mixed with tears dribbled on my bed linen. She looked up at me through tears.

'I see two of you. I see two of everything. Mamma. Will I be all right?'

Since she was a little girl, when anything happened, a scratch, a fever, a pain of any sort, with total faith she asked me: 'Will I be all right?' and waited for the verdict with an absorbed expression of complete trust.

My heart fluttered, at her reverting to childhood in a time of pain and need. She was alive, yet she might be disfigured. She could be blinded.

She moved and tried to sit. Her eyes rolled, I held out my hand in time to help her gently back to her pillow. She was suddenly sick. I took her in my arms, trying not to touch her skin, giving the others quick instructions.

In the hurry of packing a small bag after arriving yesterday from the United States, I had forgotten to take my first-aid kit. There was no Mercurochrome. No antibiotic cream. Nothing really but a harsh disinfectant that I knew would cook the skin badly, and butterfly plaster stitches that I knew I had to find the courage to use. Boiled water then; ice. The delicate skin below her eye should not be disturbed. Radiocall the hospital. Ask their advice. No question of having enough daylight hours to be able to make it back to Nairobi. And tomorrow would be too late for stitches.

'I can't see. I see double. Two of you. Two beds. Two Aidans. I feel sick. Sorry.'

She was sick again.

'Mamma. Mamma, will I be all right? Will I be able to see again?'

There had been narrow escapes before. The time she had been attacked by killer bees. The time her horse had bolted when she was pursuing zebra, and she had been catapulted into the rocks.

The time that she, still a toddler, had refused to board the car already packed to go back to Nairobi from Laikipia. She had stopped in her tracks, unusually stubborn, ready to throw an uncharacteristic tantrum. It was Wanjiru who had told me to humour her.

'*Nyawera,*' she had said, calling me by my Kikuyu nickname, '*Makena najua. Kama yeye ataki hio ngari, lazima iko na kwa sababu.*' 'You, The One Who Works Hard, The One Who Smiles knows. If she does not like that car, there is a reason.'

Reluctantly, I had moved all the luggage to a land cruiser. I felt the absurdity of the situation, but Sveva's and Wanjiru's coalition was too powerful to fight.

Next day, when our driver Karanja drove a friend who had been staying with me in Laikipia to Nairobi in the other car, over the Kabete escarpment it lost a wheel which we did not know was loose. Out of control, the car flew over the bank, overturned and crashed, and my guest fractured her spine. I drive much faster than Karanja. The thought of what could have happened had we been in that car, dried out my saliva. Inscrutably, with the African acceptance of the improbable, Wanjiru had nodded.

'*Asante Makena,*' was all she said. 'Thank you, The One Who Smiles.'

There had been the time that, during an evening walk with the dogs at Paolo's dam beyond Kuti, we had wandered without realizing it into the middle of a herd of elephants. They had been so silent that in the twilight I had not seen them. Unknowingly we had ended up almost between the legs of a towering matriarch invisible in the dusk. A sudden blare of trumpeting coming from a yard or so above us, ivory tusks protruding, a huge head shaking, flared ears. And in an instant I had gathered up Sveva, a child of four or so, and had run at an angle towards the elephant, in the direction of the wind, to get out of the line of scent as soon as possible. My heart in my mouth, legs scratched by thorns, trying to avoid the remaining herd, only after a while did I realize that Sveva, bobbing up in my arms, was waving at the elephant, giggling.

And there had been, of course, the time, soon after Emanuele's death, when a little friend had pushed her by mistake into the swimming pool at Kuti, and run away in fear. She had sunk to the bottom, a baby who could not yet swim, but inexplicably she had managed to climb out alone, soaked, but safe. There had been other stories. Accident prone, as Paolo had been, there was however something unknown – or someone invisible – which saved her in potentially tragic situations, some benign force that had so far rescued her.

Now the story came amongst broken sobs, and it chilled me.

They had sailed out in the choppy sea with the irresponsibility of youth, without life jackets. Sveva thought it would be easy, thought her friend would know how. The current of the receding tide took the small craft further and further from the shore. There were no other boats to be seen. The place was about to be closed for the Long Rains season, and all craft had been taken ashore. Somehow I had been spared the horror of watching, useless witness, impotent to intervene. That would have maddened me. Sveva tried to grasp the slippery side of the boat to keep her balance. Then the wind caught the sail violently, the boom turned, hit her across the face, flung her into the water. The boat capsized. Sky the colour of fright swirled above her eyes.

Stunned and unconscious she went under water, heavy as lead, but inexplicably she did not sink and bobbed up again. The cold made her regain consciousness and miraculously she managed to keep afloat, pain burning her face, eyes blinded with blood and salty water, waves higher and higher, and no life jacket. The aquatic Paolo streak prevailed in her as it had long ago, when she was pushed into the pool in Laikipia.

Yet again, when she could easily have drowned, she did not. The pain and the chill kept her conscious and she managed to stay afloat. Then the boy grasped a lock of her hair, floating like a bronze seaweed amongst the driftwood, and lifted her on the now righted boat, where she embraced the mast. Painstakingly, they reached the shore when Aidan, who was working at his papers in the cottage, was alerted by a boat attendant.

The afternoon drew on. Through the crackling of static, on the radio, the nurse in charge at the Casualty department at Nairobi Hospital gave us instructions, reassured us. We were to keep Sveva cool, quiet; give her pain-killers. Tomorrow we would leave at sunrise for Nairobi, and the hospital. But for today, only I could help.

I kept vigil over her agitated sleep, anxiously drying out oozing serum mixed with blood and praying that her face would be spared. When darkness fell I knew that the time had come for me to stitch the cut. It was drying out open, a gash shaped like a fan, and, unless I acted now, it would stretch her smooth skin and leave a nasty scar. Aidan held high the hurricane lamp and the young man a torch while I, conscious that it was up to me to keep her beauty, steadying my hands and biting my lips, applied the stitches to my girl's face one by one. From the swelling and pain I knew that her cheekbone was fractured. Perhaps the eye was affected. I finished finally. But she would not die.

Two years have gone by now and she is unblemished. The eye has finally recovered, and even if her eyesight has been slightly affected, the bone has mended and only a fine crescent-shaped line, translucent like a new moon, remains below her left eye to remind us of that afternoon in Kiwayu.

The doctor next day commented that the stitches were applied perfectly; mother concern and love had guided my hands. But Sveva and I knew that it is the spirit friends who had helped again. Until next time.

Next time was a couple of years later, here in Laikipia, while she was driving with her friend Tim in a wave of music along the short cut between the Big Dam and the Wilderness Centre.

My large land cruiser climbed a steep slope too fast, landed sideways on the opposite downhill side and bounced up.

She hit the brakes, lost control, and the car rolled. She banged her head on the ceiling, and was catapulted like a rag doll, stunned, unconscious, to the back of the vehicle. Driverless, the car kept going at full speed, rolled once more and landed on its roof. The front windscreen hit a jagged stump and exploded in a thousand crystals.

Shahar, the agriculturalist from Israel who had at that time just joined our team, was riding his bike towards the Centre, when he saw the car: upside down, the engine still running, music still pouring out and no one inside. The stump of an acacia branch, pushed against the front seat, would have beheaded her, had she still been in the driver's seat.

The car roof had caved in. A death trap.

The tracks of naked feet were visible in the red dust, and Shahar followed them.

She was walking in a daze, holding Tim's hand. She had crawled out of the window, extracted and comforted him. His collar bone was broken.

She did not have a single scratch.

Then there was the night in Nairobi a couple of years ago. It was her last day of school in Kenya before leaving to study in Europe. She had gone out to the end of year party.

I went to bed early and woke up suddenly in the middle of the night.

There had been a noise.

It had been raining earlier on but it had stopped now. Water dripped from the guttering. The dogs were silent. Someone was knocking, urgently, at the downstairs door.

'Mummy, open the door.'

I shook off the sleep.

'What happened to your keys . . . it's four in the morning . . .'

'Mummy, we've been robbed. They took everything. Open the door. I'm all right.'

Soaked in rain, her long hair damp on her shoulders, Sveva stood with her friend Tim on my doorstep. Over a steaming mug of tea the story unfolded.

Their car, moving slowly in the heavy rain, had been ambushed up the hill, in Forest Road. A *panga* had gone through the half-lowered window, Tim had been yanked from the car and made to kneel on the grass at the side of the road. His face was pushed into a muddy puddle, a foot placed on his back to hold him down, while a bandit searched and robbed him. Sveva had crouched inside, invisible in the dark, protected by the fogged windows. With self-preservation she had removed from her neck her golden chain with the pendant of the two trees logo I had given her as a present. She hid it between her breasts and she got out. Caught unawares, the bandits turned. One leaped towards her, pushed her, took her purse, grabbed her blouse, began to undress her. Time stood still.

Smiling unexpectedly at angry, tense and frightened people disarms them, takes the wind from their sails, acts as a balm. It is often infallible and even this time it worked.

Standing in the night rain in deserted Forest Road, Makena bit her lip, swallowed her fear, smiled at the robbers and joked with them in her fluent, gracious Swahili.

Could they not see that they were young, and there was nothing much they could get out of them, they had a little money, here it was, they could take her watch, here it is, take it, a good watch from Europe, and her necklace with blue lapis lazuli, it was real silver – you can always get another watch, another necklace – could they not see it was cold and they were wet, all they wanted to do was to go home?

The bandits were caught unprepared. They knew how to deal with rage, with fear, with aggression. Her calm, her words, her bright smile disarmed them.

Once again, the silent star of Makena's luck prevailed; another escape, as for Paolo, long ago. The page was turned, the time for violence gone, the robbers disappeared into the fog, like a nightmare at dawn.

The rain kept falling on the deserted road.

Return to Moyale

Our camels sniff the evening, and are glad.
FLETCHER, *The Golden Road to Samarkand*

The aeroplane circled low over the tin roofs of the small frontier town built in Arabic style, headed towards the airstrip, and landed in a cloud of dust, hardly disturbing the camels. They looked up from the thorn-shrubs they had been browsing, totally unaffected. I noted how very white they were, like ghosts of camels.

It was a morning of September 1997. We had come to Moyale, in the extreme north of Kenya, on the frontier with Ethiopia. I was curious to return, for happy and poignant memories tied me to this remote and god-forsaken land.

The first time I visited Moyale was with Aidan, in 1990, in preparation for a camel trek. The purpose had been to make arrangements to purchase, and, the following year, to bring a herd of breeding Somali camels to the Laikipia region.

The itinerary would cover just over 300 miles across Kenya's Northern Frontier district, from Moyale to a place called Lolokwe in Samburu district, next to the *lugga* of Il baa Okut, and eventually to Ol Penguan, Aidan's ranch in the savannah of the western Laikipia plateau.

For the first 200 miles, until the wells of Koiya, there would be

247

no water. To take advantage of the cooler temperature of the night, the march was planned around the time of the full moon in September 1991. There were reasons for such a feat of endurance.

Aidan's passion for and understanding of camels was deep rooted. The original African livestock, adapted to the harshest conditions, to desolate places with scarce water, sparse forage and scorching heat, camels are far kinder to the ecology of the land than cattle. Their soft, padded feet do not erode the soil; they browse discriminately from tall shrubs beyond the reach of goats, thus never overgrazing the meagre pasture of Africa's most arid regions. Their dense, slightly sour milk is extremely nutritious and constitutes the largest part of the diet of most of the nomadic tribes of northern Kenya, whose lives completely depend on these animals. Unlike cattle, camels do not need to drink daily; they cover vast distances effortlessly, carry loads and are immensely adaptable.

It is clearly desirable to introduce camels in areas of Africa where they are not indigenous, but where conditions are suitable, as an environmentally friendly alternative to conventional livestock. They are also an attractive and decorative substitute for horses for tourist treks.

But a true passion is seldom born of reasoning; what most appealed to Aidan's noble, nomadic soul was the romantic aura surrounding these haughty, stately creatures, 'the ships of the desert', dignified, aristocratic, resilient, indispensable companions to any explorer of the Africa Unknown.

Another of Aidan's deep interests was collecting rare and often undescribed plant specimens, particularly bizarre subspecies of succulents like euphorbia, caraluma, aloe or aechinopsis, which still grow undiscovered in isolated, almost inaccessible gorges and peaks, weeks away from any beaten track. Most of his botanical expeditions would have been impossible without his camels.

When breeding livestock it is essential to periodically introduce new blood. This was the main reason for the scheme. The other one was, of course, the challenge of the difficulty, the beauty of the land, the danger of going through uncharted territory with just the basic essentials.

Purchasing dozens of camels, and in such a remote location, is a complex operation that cannot be accomplished in a few days, and much thought and planning went into organizing the expedition. Aidan was familiar with the procedure and knew the people without whom such a plan would have been impossible.

So I flew to Moyale with him, to meet his friend Haji Roba, an old dignitary quite prominent in that small Muslim community and a crucial figure in his plan. He would be entrusted with a substantial amount of cash. He would keep it in his rusty old safe, in the back room, cool with thick, whitewashed walls, separated from his front parlour by a yellow muslin curtain embroidered with red roses. He would hand out money over the following weeks and months to Aidan's man Ibrahim Mohamed, who was already on his way by road and would be based in Moyale for however long it would take to purchase a hundred camels.

Camels were brought to the market in Moyale from across the border, tall Ethiopian camels of illustrious breeding, the females good for milking, the males bred to walk tirelessly and carry loads. One day there would be five, another perhaps ten, and some other day none at all good enough to be purchased. They would be brought after long bargaining – the dealers sitting on mats strewn on the dusty gravel in the shade of an old tamarind, next to the *sokho* and not far from the mosque, just out of reach of the omnipresent goats – after drinking endless cups of tea with cardamom and comforting glasses of smoky-sweet *susha* curd.

It would be a man's world, in Muslim tradition, a world of business and trade, of caravans and mules, and talk of *shiftah*.

Shiftah.

Shiftah is the Somali name for 'bandits'. They are highwaymen belonging to various ethnic clans, who continuously encroach from Ethiopia and Somalia to raid the borders of Kenya. They take advantage of the mountainous territory, inaccessible to administration, where they can easily disappear and escape capture.

Not a day would go by without *shiftah*-related incidents being reported: sometimes it would be only stealing a mob of camels, never without a lot of shooting, but in most cases there were serious casualties.

Shiftah would kill for livestock, but also for clothes, for shoes, for a watch, for the sake of killing. *Shiftah* were wild, brutal, unpredictable and without mercy; often high on local drugs, they would raid *manyattas*, beat and rape women, slaughter children. They were not brave and gallant warriors, but bands of desperados with no honour, who, strong in numbers and weapons – they never attacked alone – would pounce on a solitary traveller and cut his throat for a few shillings' loot, or just because he belonged to the wrong clan. They would gallop in on fast horses, shouting and spraying bullets. *Shiftah* were no joke, and the only way to win was to stay clear of them.

Vehicles were no longer allowed to go on the murram road from Isiolo to Moyale without an armed police escort.

I remembered that first day in Moyale, and the secluded courtyard of our host, his lavish, generous hospitality, the shy beautiful little girl in the purple veil – his younger daughter, whose name was Rehema – slim and charming, with liquid, intelligent eyes, still little more than a child, who held a copy of the Koran, written in Arabic. She greeted me first by kissing my hand, then she moved to Aidan and lowered her head so that he could place his hand in sign of blessing; and when he did, she turned gently and gracefully kissed his hand before running back to the dark doorway, amongst the other scuttling small children. I noticed that now and again she raised her hand to her ear, twisting her neck, as if in pain.

To enter Haji Roba's house we climbed tall steps. In the shady room, we were greeted by some of his elder sons and, amid exclamations of welcome and commiseration for our long journey, made to sit on a long sofa upholstered in brown synthetic leather, before which was a coffee table covered in lace and adorned with a colourful arrangement of plastic flowers. From a charcoal brazier a pleasant resinous smoke of incense rose to the pale blue ceiling.

Soon the room was filled with other young men, other sons, and finally Mumina, his second wife, who was pregnant again, peeped from the door.

In the room with yellow cotton curtains, bright with bead-embroidered flowers, I waited for the time to pass, while the men spoke of their business.

Then it was time for breakfast – although it was more like lunch-time – and they showed me to a Turkish-type toilet built next to the kitchen where the women, squatting on the beaten earth floor, were deftly cooking delicious-smelling food on a small fire over three stones and singing.

A basin was brought, and a towel and a piece of soap, and we all washed our hands. Then platter after platter of food came, enamel dishes heaped with rice and goat stew and roast goat and goat entrails fried with onion; and *chapati* and lentils and banana fritters and sour camel milk with sugar and cardamom.

Ravenous, intrigued by foreign flavours, I ate more than I should have. Then Mumina came in and shyly embraced me, kissing me twice as is customary amongst Muslim women friends, and offered me a parcel, in which a fine Somali shawl was wrapped, red and yellow and white, woven of a very fine muslin. I draped it around my head and made a mental note to send back a present as soon as I could.

Haji Roba had organized for us to go into Ethiopia later for lunch, to meet a man who had heard of our arrival and wanted to meet Aidan. We boarded a rickety old Land Rover, borrowed from some merchant neighbour, and crossed the border without passport or any other document.

Everyone seemed to know Haji Roba; they waved at him from the side of the road and, to my amazement, quite a few people in that unfamiliar land greeted Aidan like an old friend.

We drove to a dilapidated hotel, squalidly decorated with rickety, cheap furniture, where we sat waiting for the man. He was the local government representative, a type of warlord, quite high up in the country official politics. His name was Godana.

I looked at the hotel with curiosity since it was here, seven years earlier, that Aidan had been kept prisoner for a month, when, on a botanical expedition for the herbarium, having been dropped alone in the middle of nowhere by an aeroplane, he had been found by the local Garreh troops. Wandering the countryside with maps and compass in his rucksack, but no documents, he had been mistaken, with some logic, for a spy: no one could quite believe that all he wanted was to climb some unexplored mountain and collect rare plant specimens. *Wasungu* are all crazy.

Barely sustained by a monotonous diet of overcooked spaghetti of infamous quality, grainy coffee (dubious inheritance of Italian colonial days) and small bananas from a nearby grove, he was left there to ponder on his destiny, the door guarded by a sentinel who went with him even to the loo, until his rescue was organized.

Finally, the Ethiopian man arrived; with him came a polite, shy youth with eager eyes and sleeked-back hair, his son, Wako. So it transpired that university in the States was the young man's dream, an impossible dream, like journeying to a distant galaxy and we were the one bridge to that unknown. Feeling strongly for the boy, in subsequent months I did all I could to help. Years later the phone rang in Nairobi from the States, and there was Wako's jubilant voice, from far away, he who had made it after all.

We returned, making plans for the journey. Once the camels were purchased, we would fly in with Aidan's brother-in-law's airplane, a large, wonderfully old-fashioned de Havilland Beaver which could carry our bulky load. Our ambitious itinerary was dangerous, for the area was known to be riddled with *shiftah* bandits from Ethiopia. A large caravan with two Europeans would attract much attention, and we would have to avoid all known trails, and the crowded water holes for the first and major part of our trek to avoid being ambushed. The march was planned across dry savannah, deserts, lava stretches and seasonal swamps, with no roads or villages, and reachable only on foot. There would be no way of communicating once we had started our safari. The time of satellite telephones was yet to come. I would bring as comprehensive a first-aid kit as I could, and would hope for the best. With no refrigeration, there was no point in taking snake serum.

We would carry all the water we needed for drinking and cooking; we would wash only after reaching the wells of Koiya. Aidan would bring his rifle and the best camel men from his ranch: the two Ahmeds, distinguished with the nicknames of Nyukundu (red) and Nyeusi (black). I would bring two of my security rangers, with their guns, for protection. I would choose one who could speak the Adjuran language. We would employ some more camel handlers in Moyale.

Their leader would be Osman.

I heard that day, for the first time, his amazing story.

The Story of Osman

But one man loved the pilgrim soul in you,
And love the sorrows of your changing face.
                    W. B. YEATS, *When You Are Old*

There was a little man who looked after camels, and his name was Osman Nguyu Dupa.

He was of slight build, with frail, bird-like bones but knobbly knees and large feet, like the animals he tended. He wore a long, cotton checkered *kikoi* and a loose turban, tied a few times around his grey hair, in the Boran style. He came from the Northern Frontier, close to Ethiopia, the land of sands and deserts, of lava rocks and unforgiving sun. His skin was dark and shiny, stretched over wide and prominent cheekbones on a haggard thin face with chiselled features, like the naïve black Christs painted on Coptic icons. His eyes were bright, mobile, used to scanning elusive horizons of trembling mirages for a sign of water, which means life; of endless dunes, for a sudden movement which may mean *shiftah*, and death.

His existence was of a biblical simplicity, timed to the camels' life and to their needs. Sunrise and sunset prayers began and ended every one of his days. He looked after the camels with brisk efficiency, singing to them, calling and herding them with his camel stick, walking tirelessly amongst them, behind them and ahead of them, by the end of the day covering easily ten times more miles than they had. He worked for Aidan as his head camel herdman.

Over the years they had shared many adventures. Safaris in the northern lands, walking for months through unexplored territories. They had shared star-studded skies in windy nights around the same campfire, and drunk from the same wells. Together they had risked their lives. Discovered once wandering illegally well into Ethiopia with their camels, Osman and Aidan had been taken prisoners by a local Garreh tribe, and held in their *manyatta* as spies. The local Ethiopian garrison, alerted by a runner, reached them on foot days later. After a summary trial under the scanty shade of the sparse thorn trees, they were finally freed. Small things, nuances, were responsible for their release: the women, maintained Aidan, had been instrumental in such leniency, appreciative of Aidan's help in lending the camels to help them carry their water from miles away; the men's curiosity about Aidan's rifle, and their gratitude for the penicillin he offered to cure the rampant venereal diseases.

A deep mutual respect had grown over the years between Osman and Aidan, rooted in the common love and understanding of camels, the invaluable creatures without which no desert can be crossed on foot.

It was therefore with great dismay that Aidan realized one day that something was very wrong with Osman. He could not touch food, complained of a terrible pain in his stomach and seemed to lose weight almost by the minute. Finally, when he was reduced to a light bundle of skin and bones, Aidan had carried him in his arms to his airplane, and had flown him off to the Italian missionary doctors at the hospital in Wamba.

Wamba is in the territory of the Samburu, a landmark of eroded dry red earth scorched for miles around this extraordinary hospital in the middle of nowhere, where dedicated doctors minister daily to all imaginable diseases.

In the X-ray slide, an ugly, opaque mass occupied most of Osman's stomach. The tests confirmed that a malignant cancer in its final stages had expanded to extreme conditions, and that operating would be impossible. The kindest thing was for Osman to go back to his family and there to die there in peace.

One sunny morning, in a deep blue sky, Aidan flew him off to faraway Moyale, a corpse almost, and left him there in the shade of one of the rare thorn trees, his wife and children gathered around him in a silent circle, mourning already the inevitable loss. Alone over lava deserts, flying the long hours back to his ranch, Aidan felt with a heavy heart that he had seen the last of Osman.

Then, soon afterwards, Aidan heard news of a famous faith-healer who came periodically to visit terminally ill people in Gilgil, and of the amazing results of his ministrations.

In this land of the inexplicable, where most believe in magic, and where anything can happen, Aidan decided to give his old friend another chance. He booked an appointment, jumped into his plane, flew three hours due north, located the camel camp and Osman's *boma* from the air, circled twice and landed on the track in a cloud of white dust. Then, leaving behind a small bewildered crowd of people, camels and goats looking up in amazement from the remote strip of sand, he brought Osman in hope and glory back to Gilgil.

Again, he carried in his strong good arms his dying friend wrapped in a threadbare blanket.

It was a quiet, pleasant room with a few stools, a house plant or two, a table with a jug of milk, a pot of tea and some pink sugar-coated cake set on a tray. In the middle were placed a bench and a narrow bed. And there he laid him gently.

For many of the Africans, still unpolluted by western complexities, quaintly coexist, without conflicting, a total respect for their tribal witchcraft, and an almost unlimited faith in the capacity of European medicine to achieve the impossible.

Moreover Osman was a Muslim, a man of God, and there was in him a complete belief in Aidan's omnipotence: his companion of many adventures, the man who walked alone, who knew camels and could shoot straight, the kind, serious giant who was known to master the bird of metal with unparalleled skill, and had appeared from nowhere in the merciless northern sky to save him, was surrounded in people's tales, and in his heart, by an ineffable mystique. It was perfectly easy for him to accept that Aidan and the healer of the *musungu* God by the white hands would help him where everything else seemed to have failed. And it is on unquestioning trust and confidence that faith-healing is based.

They calmly explained to Osman what would happen. The doctor would put his hands on Osman's stomach, gently probing the swollen mass. Perhaps some heat would be felt. Perhaps nothing. Perhaps, if God was looking, the pain would ease, the evil ball would dissolve and Osman, if God wanted, would survive. *Inshallah.*

He was so weak that he was past caring. For weeks now his only food had been a few sips of camel milk, swallowed with difficulty. For months his constant companion had been an agony of pain. He had come to wish for an end to his misery, whatever this end might be. Death, the great forgiver, shrouded in white veils, would have been merciful. He closed his eyes, exhausted, dizzy with throbbing pain, ready for miracles.

The healing hands lightly touched Osman's tummy and a dry heat radiated all through it. How long it lasted he could not tell later; perhaps half an hour? But when the holy hands were lifted and the weary doctor wiped his forehead, Osman opened his eyes. He looked up and around, scanning the room with a new interest, focused hungrily on the dish of tea and the pink cake, pointed a bony finger and looked at Aidan.

'*Chai. Mimi nataka chai na sukari minghi. Na kipande ya keki, kubwa sana. Kwisha pona. Mimi ni sawa sasa. Allahu akbar.*' 'Tea.

I want a cup of tea with lots of sugar; and a big piece of cake. I am cured. Allah is great.'

He spoke with no surprise, accepting the inevitability of the miracle, Aidan's power, and the doctor's magic all at once.

The conventional doctors in Wamba were disconcerted. The foul cancer mass was no longer visible. The X-rays were clear. Incredible as that might be, Osman had completely recovered.

They flew back to Moyale once more, this strange couple, the tall silent man in shorts and the small African in his *shuka*, landing again in the dust of the livestock trail. Again the children ran towards the shadow of the aeroplane, and, huddled together around the veiled tall woman who was their mother, waited in anxious silence for the plane to come to a halt skidding in the dust.

Even the cicadas had stopped singing.

The door opened, and in that silence Osman jumped out alone.

A collective intake of breath ran through the small crowd like the shiver of a sudden wind. Then the joyous whining began, the suffocated giggles, the lilting sing-song, the querulous talking all together.

'Get your strength back,' said Aidan in his farewell, 'and when you are fit again I'll send for you, and you must be ready. We will buy new tall Dogadia and Adjuran camels at the market in Moyale, and when we have enough I will return. You will come back with me to herd them. We'll walk them back together.'

As it happened, I walked them back too.

The Story of the Abagatha

He was a man, take him for all in all,
I shall not look upon his like again.

SHAKESPEARE, Hamlet

The pilot nodded to me and to John Wachira, the doctor from
AMREF* with whom I was coming to visit the hospital.

'Here we are. I'll pick you up on Monday.'

So there I was, back in Moyale, looking forward to the adventure. A different type of adventure from the last time I was here.
I had agreed to help the Flying Doctors of Africa to raise awareness for their Outreach programme, and this was a fact-finding
mission to familiarize myself with what was happening in the
remote corners of Kenya where they fly in regularly to administer
all types of specialized surgery and medical care, a daring, complex, romantic and desperately needed undertaking. I would
write a story for them. I would stay in a hotel in the Ethiopian
part of the town, and spend the day at the hospital, where a room
had been set apart for me, so that I could write in peace.

I stepped out of the plane and blinked in the harsh sunlight. The
airstrip stretched dusty and unvisited; a shredded windsock hung
limp for lack of any breeze. A few long-robed women, their heads
covered in bright veils, stood on one side. Children came run-

*African Medical Research Foundation.

ning, materializing from everywhere and nowhere. A Land Rover
approached and people got out to greet us. Amongst them, I
recognized Haji Roba. Tall and distinguished, a white Muslim cap
on his head, khaki trousers, a coat and his stick. He came towards
me smiling, holding my proffered hand in both of his, exuding
genuine pleasure to see me. Having heard of my arrival, he
wanted to find out if there was anything I needed. He asked me
to go and have breakfast at his house next day, see again his wife,
his children.

We drove off.

The town was as busy and dusty as I remembered, crowded
with people moving in all directions and talking in excited, gesticulating groups. Slender ladies, clad in bracelets and muslin veils
of red and blue and white, moved around, effortlessly balancing
large bundles and enormous gourds on their heads. Goats for-
aged on heaps of old cabbage. Donkeys dozed wearily in the sun,
and tall camels, tended by noisy leaders, ambled to the *sokho*
carrying loads of merchandise from the Ethiopian side of the
town. Dressed in flowing *kikois*, their bearded heads wrapped in
shammahs, dignitaries conferred below the tamarind trees. The
hot, still air smelled of dust and spices.

My return to Moyale was studded with novelty and unforgettable incidents, glimmers of another simpler and ancient world.
I shall remember the hospital where young, bright Doctor
Mohamed attended to a hundred horrific emergencies with calm
simplicity; the hopeful patients crowding the corridors like biblical figures in their casually elegant robes. The hotel where I slept
in the Ethiopian town, a paw paw tree growing just outside my
door, and huge cockroaches hiding under my bed. The eyes of
the bossy little girl in charge of the shop where I purchased an
embroidered white *shuka*, her sudden disarming smile when she
said farewell in Italian. The young beggar with the gangrenous
leg. The smell of the first sudden shower of rain on the dust, and
the colourful market where, after much bargaining, I bought

Sveva a red shawl. The woman pounding coffee grains with a long stick in a wooden pot, singing. The flavour of *gored gored,* the spiced stews, served in an array of *sufurias* set on the hospital floor outside the theatre at lunch time, when I joined the weary surgeon for a short break between operations. The muezzin's song from the mosque, merging with the prayers of the Coptic priest from the orthodox church.

But, mostly, the brief encounter with the Abagatha of the Borana, and the story of the *muganga* who cured Rehema.

Haji Roba disclosed that there was great anticipation amongst the Boran clan – a major proportion of the town's inhabitants – as news had come of the visit next day of their Abagatha. The Abagatha is the Boran's traditional leader, something between a tribal king and an elected president, and the highest authority of this noble tribe. He was sick, had heard of the flying doctor who had come to Moyale, and was being driven three hundred miles from their tribal domains to come and see him.

'Being driven?'

'Yes, the Ethiopian Government will give him a car,' revealed Haji Roba. 'The Abagatha of the Boran does not have his own vehicle. Boran ride horses.'

He explained to me about the Boran clan, to which he belonged. It is the largest Ethiopian clan, with several millions of people. They stretch from Ethiopia to Kenya and though they are citizens of the countries where they live, their ultimate loyalty is to their clan leader, the Abagatha. They follow their tribal law strictly, a fact acknowledged since the times of the British Colony, and accepted by the governments of the two countries.

The Borana are generally peaceful, but immensely powerful. Their traditions are deeply rooted. Their Abagatha is chosen from one of the families which have already produced other Abagatha, usually but not necessarily when still a child. He undergoes a complex education in their law, customs and secret, sacred ceremonies. He is recognized as the religious and political leader at the same time. He will be the Abagatha for eight years, when another trained youth will take his place.

Although most of them are now Muslim, traditionally Borana are neither Muslim nor Christian. They pray to God in the morning and to God in the evening, and have their rituals, but no specific place of worship. They can take as many wives as they choose. They are horsemen, warriors, livestock people, brave, gallant. Their Abagatha moves from camp to camp and so gets close to his tribe and their problems. His word is law, but he is known to be fair. He travels around with a retinue of advisers.

Naturally I was intrigued and extremely eager to meet this charismatic and mysterious character.

When we arrived at the hospital on the second day I knew instantly that the Abagatha was already there. An official-looking Land Rover with Ethiopian number plates was parked below a dusty palm tree, and groups of people stood expectantly at corners. From the silence and austerity prevailing I imagined that the patient would be a grand, dignified old man.

I was unprepared for the young person in long white robes and sandals, a type of round, rigid turban on his proud, handsome head with chiselled features, adorned by a lustrous beard. Obviously in pain, he sat on a bench and looked up at me searchingly with burning, intelligent black eyes. The ailment appeared to be an old calcified hydatic cyst in his liver, caused by a dangerous local parasite that becomes embedded in the body and grows into grotesque swellings, infiltrating every organ until the patient dies.

The medicines were not available at the hospital, and I offered to get them in Nairobi and dispatch them in the army aeroplane which brings supplies to Moyale every Wednesday. On learning this, the Abagatha sent an old man to tell me that he wanted to thank me personally. Soon he entered the small office they had given me, followed by all his escort, but without any special ceremony. He sat there on a narrow bench, communicating through an interpreter. He asked where I came from, and what was my occupation; he assumed I was a doctor, too, and looked at my computer with curiosity. On learning I was a farmer, he wanted to know if the rains had been good, if the grass was long and the grazing abundant, my camels fat and my cattle healthy; how many children I had. His penetrating eyes never left my face while the interpreter translated from Borana to Swahili. When he heard of Emanuele's death from snakebite he winced once.

From his concentrated gaze I could see something was puzzling him, and finally the most important question came, and it startled me. He wanted to know about my hair, which is quite long and streaked with blond. He explained that he had never seen hair which looked like a lion's mane before. Was it real hair? He moved a bony hand, its long sensitive fingers adorned by silver rings, to touch a lock on my forehead, and laughed suddenly, with white gleaming teeth like fresh matched almonds. I realized that I was possibly the first white woman he had ever seen, and certainly spoken to.

So it was that I was invited by the Abagatha of the Borana to be his guest, to stay at his camps, and share his food at his campfire, on the hills over the desert from where they can see the blue plains, where the nights are chilly and the days hot and dry; where they keep hundreds of horses, sleep in tents made of hides and time does not matter. I accepted immediately, of course. We shall decide on a date and this will be, I know, the ultimate adventure and another story to tell.

Haji Roba came back later that day as promised, to bring me first to the market – where, fiercely and jovially, he did all the bargaining on my behalf – and then to his house for breakfast.

I had brought a present for his wife, Mumina, an ornate golden thermos for her *chai,* incense sticks and a large box of sweets for his children.

In the internal courtyard I met his two new boys, all dressed in their lacy best clothes, who ran to greet me with lowered heads so I could put my hand on them in blessing, and a baby girl of a few months carried by a maid, all born to Mumina in the last six years. Mumina was older, rounder, but still graciously, maturely handsome, and the liquid, kind eyes glinted in welcome as she embraced me.

The house had not changed much. The curtains were now a pale green and the sofa covered with a synthetic type of leopard print. The walls were freshly varnished, deep green and white, and I noticed with surprise that the gold pendulum clock I had donated to one of Mumina's baby boys six years earlier, at the time of the camel trek, was still working, taking pride of place in the middle of a wall.

For breakfast I was served roast and boiled goat, fried rice, *chapati* and *samousas* and a good sponge cake with tea.

Haji Roba sat with me, eating sparsely with his hands, urging me to help myself, and narrating all the news of the past few years, slowly, with a wealth of details.

His adult sons had died. One had been ambushed by *shiftah* on the road to Isiolo. Everyone in the lorry was shot, including the armed escort. The other two were both wiped out during the epidemic of *dengi* fever in 1994.

As he spoke of heart-rending sorrows, of graves dug in the sand amongst the grey salt bushes, guarded by marabou storks, his genial face bore a mask of melancholy, and once again I marvelled at the acceptance of death of the African people.

And what became of Rehema? I had a vivid memory of the beautiful little girl who had served our food and could read the Koran, clearly a favourite child.

His face lit up and the wrinkles disappeared. Rehema was at a boarding school in Meru. She was doing well with her studies, and she had recovered physically, cured in the most curious of ways of the recurring ailment that had defeated all doctors, a chronic infection in her ear, which – as I would remember – had kept her in agony since childhood. Did I not know the story?

The Story of Rehema

For my part, I have ever believed, and I know that there are
witches.

Sir Thomas Browne, *Oh Altitudo!*

After the last doctor in Nairobi, the Indian specialist at the
Aga Khan Hospital, declared that there was nothing he could
do to cure Rehema's ear, Haji Roba heard news of a *muganga*
from Ethiopia who could heal practically anything.

He was a wild fellow from the tribe of the Surma, uncouth
and untamed, with a savage reputation. He lived in primitive
conditions in a tent of uncured hides set in the bush, and never
wore clothes, an incomprehensible custom for a conservative
Muslim.

After much pondering, they finally sent for him, and when
Mumina saw this naked man with oiled limbs in her clean parlour, she covered her face with her shawl, and immediately dis-
patched a servant to buy some clothes for the *muganga* to wear
before she could meet him again.

The man took the clothes, spat on the floor and wrapped them
around his dreadlocks in a bulky turban, remaining naked, apart
from the tiniest shred of material covering his genitals.

A reluctant, terrified Rehema was eventually brought to the
room where the *muganga* crouched in a corner like an animal.
He approached her, crawling, breathing noisily like a wild creature. He smelled her body from head to toe, stopping finally at

the back of her head, still sniffing. Then he extracted some short sticks from his leather pouch and confirmed his diagnosis by passing them over her head and limbs. When the sticks began to vibrate, seemingly on their own, near her neck, behind her ear, there was no doubt. With a small knife the man made an incision behind Rehema's right ear which was, indeed, the one which hurt. Then he applied his big lips to the incision and started sucking. Poor, petrified Rehema cried and whimpered, trying to struggle away, but, undaunted, the man sucked and sucked, until – lo and behold – two squirming, white, fat worms wriggled out from the incision along with two small cockroaches which the man spat out in triumph, showing them to the revolted audience with a devilish grin.

I was disgusted. Haji Roba went on.

A goat was brought into the yard and slaughtered, her dark blood spilled on the dust like a prayer, and collected in a small gourd cured with ashes. The man smeared the still hot blood on his hair and face, and liberally doused the barely breathing Rehema. She was then, mercifully, sent to bed, where she tossed and cried for a few hours before falling into a fitful sleep, populated by the monsters of her nightmare.

When she woke up she was washed, more blood was smudged all over her head, and the foul man again applied his greedy, obscene lips to her tender neck.

By now the man's mouth had been inspected by the entire family, scrubbed and rinsed with a toothbrush to ensure there could be no trick, until they were satisfied that it was empty. So, when another worm appeared and three more cockroaches were spat out in the basin, everyone caught their breath and prayed to Allah in awe and revulsion. Rehema was again sent to her room to sleep.

When she woke up next morning she was fresh and rested, the earache had gone and it never came back.

Haji Roba asked the *muganga* his price: ten Ethiopian birrs was all he asked, less than one English pound.

The man eventually returned to his wilderness, but not before Mumina's mother, an aunt and an array of sufferers had emerged to be cured. He extracted some hot, wet stones from a deformed foot, and some small snakes from someone's stomach, and disappeared in glory, naked as he had come.

Haji Roba adjusted his embroidered cap and smiled his crafty and humorous grin, and I never quite knew if he, a true Muslim, really believed the pagan ritual that he had described.

'*Rehema kwisha pona,*' he beamed, shrugging. 'Rehema is cured.'

And this was all that mattered.

Haji Roba looked assessingly at me, and asked about the camels we had brought from Moyale in 1991, and how they were. Around my neck I still wore a small leather pouch he had given me all those years ago, which contained a page of the Koran, a priceless amulet to protect me during my travels. He shook his head a few times, shrugging off indulgently our European eccentricity, and told me that he had been quite impressed when he heard that we had actually made it, crossing the deserts on foot, avoiding *shiftah,* in such harsh conditions, because I was a woman, did not look that fit, and that journey was a true hardship and a thing for men only.

The camel journey had been a feat of perseverance, with no pauses for leisure, for washing, for resting. But the exhaustion brought by the long marches in the scorching sun; the dust, heat, swollen feet, blisters, sweat, thirst and ticks; the constant watch for predators and rustlers; the fear of losing the way; the fatigue of keeping up with the brisk, unrelenting pace of the camels; the worry for their safety, health and feed; the need to reach the watering wells within the limits of their and our endurance: all this had been tempered and made tolerable by the heady elation of moving through unbelievably beautiful, wild and unvisited country, and for me, by the awareness that such a rare and unrepeatable adventure was the fulfilment of a long cherished dream. I had, in my early twenties, been crippled in a tragic car accident. An invalid for a few years, unable to walk without crutches, I had dreamed of being able to march on healthy legs over the African savannah.

It had seemed impossible. It was happening now.

I wrote a diary of the trek every day, either at night by the light of the campfire or when we outspanned in the heat of the day beneath sparse, skeletal shades. It was an exercise book bound in black cardboard, with curled-up pages discoloured by sun and rain, in which I wrote in pencil.

It was still fresh in my memory. Before returning to Moyale with the flying doctors I had read again the account of that glorious, unforgettable journey, my journal of 1991.

And here it is.

But was there only dust, at Arba Jahan?

The Camel Trek

Where my caravan has rested
Flowers I leave you on the grass.

EDWARD TESCHEMACHER, *Where my caravan has rested*

*Journal of a camel trek from Moyale to Lolokwe, via
Il baa Okut: September/October 1991*
with: *Osman Nguyu Dupa, Ibrahim Mohamed, Ahmed Salat*
(Nyukundu), *Ahmed Ibrahim Adan* (Nyeusi), *Mamhood,
Gedi, Ibrahim Ahmed, Lwokignei Meto*

DOLOLO NURE, 19 SEPTEMBER 1991
Fujiyanyota – Dololo Chaka

The De Havilland Beaver flown by Tony Dyer landed at Kuti
yesterday, at about 9.30 a.m.

We are ready.

Yesterday Aidan flew in late. He was still not completely well.
The virus which has swept through Kenya and which has affected
us both, has settled in my ear and affected his liver: what rotten
luck. Because of this we had to delay our departure a couple of
days, but we have finally resolved to go anyway, as everything is
ready, our hearts are set on the adventure ahead, and we have to
take advantage of the full moon.

We packed the night before to have all ready in the morning.
As ever, I have more food than I can fit in the camel boxes. I have
to rearrange tins and packets several times before I am satisfied,

and much has to be left behind. Shame about the salmon and artichoke hearts, but I manage to smuggle in a chocolate dessert.

I bring two of my security rangers with me, a gun each, Ibrahim Ahmed and Lwokkignei Meto. The latter is a young Turkana, while Ibrahim is of the Adjuran tribe, as are most of Aidan's camel people, and they know the main language of the areas we shall cross. We are unlikely to encounter anyone, but should we meet people we shall need interpreters.

So we take off on the bright blue morning of 19 September, bound for Moyale and our adventure.

Laikipia looks green from the air, but I know that leafy shrubs, streams and shady glades will soon become memories made of the substance of mirages.

We are bound for the desert.

We fly high over our itinerary: mighty and partially desolate from the air. Chains of mountains and dry *luggas*, open arid plains with not a sign of water, deserts, rocks of lava, and straight lines of abandoned oil-survey tracks, seemingly unending, diminished by a parched horizon of sand, dust and desiccated shrubs apparently empty of all forms of life.

Some ostriches and three oryx are all the animals I spot from the air; camels, a few head of cattle – or perhaps goats – and limitless dunes and barren sands.

Dried-out trees like bony skeletons, greyed by the sun.

Haji Roba expects us for lunch: he wants to give us a formal send-off in Muslim style. So I have to dress appropriately, with a skirt to cover my legs, while Aidan wears long trousers which had belonged to his father. I pack away my khakis and wear a new brick-red tricot ensemble, some jewels and a shawl. During the flight, engine oil spills onto me and stains my skirt.

Haji Roba is waiting at the landing strip, composed as ever, with a grin splitting his pleasant Boran face under the Muslim cap. We drive first a few miles in his rattling borrowed Land

Rover to take our men and gear to join the rest of the team and the waiting camels.

From a clump of thorny shrubs filmed with white dust in the area where he has been guarding his camels for weeks now, emerges Osman; I have heard his story and I am eager to meet him. He is an exquisite old chap with a slim, wiry body and intelligent, mobile brown eyes. Quite short, but straight, and thin as only Somali can be. He looks fit and sprightly.

It is hard to believe that a few years back, this same man, reduced to a skeleton, was dying of advanced cancer of the stomach. He was cured by a faith healer and in a few days, he was back to look after his camels. The mission doctor, who had, with his own eyes, seen in the X-rays the lacerated remains of the stomach devoured by the spreading cancer, and left the moribund man to a sure grave, was rather shaken when he heard the news.

Otherwise, nobody made much of a fuss about it. As this is Africa where one accepts the inexplicable.

I like Osman instantly. Nimble, on crooked thin legs, he comes forward to shake my hand.

'*Jambo, memsaabu.*' He greets me in a flash of white teeth. I notice, startled, that he has the deep voice of a much taller man.

His liquid eyes scrutinize me attentively and I wonder if I have passed the test. The trek we are about to embark upon is a feat of endurance not really fit for a woman, an immense challenge. Two hundred and eighty miles without any water, in *shiftah* territory. Aidan, a former cripple, and eight men, one of whom is Osman.

Osman. After his amazing recovery, Aidan tells me, Osman bought a new young wife and went off to dig for gold in the Ethiopian hills. He found an insignificant amount of gold after several months and much labour; and when the danger dawned on him that he might be forcibly recruited into fighting the Eritrean insurrection to which he saw many of his companions

being taken away in chains, he wisely went back to Moyale and
to his camels, his true passion.

'Can you help me?' He points to his head, where I now see a
long swollen cut, from which red thick blood is still oozing. He
hit it on a sharp stump, trying to restrain a camel when all of
them fled, startled by a hyena last night. Apart from two which
haven't been seen again, all the camels have been found miraculously, taking advantage of the full moon.

The blood looks startlingly bright and shiny on his close-
cropped hair, and I minister to him with hemostatic cotton from
my first-aid box.

Haji Roba offers us the usual choice of roast and stewed goat,
large bowls of steaming rice and meat, dates, *chapati*, bananas in
chunks and, to drink, *susha*, the smoky yoghourt made with camel's milk, spices and sugar, cardamom tea and sodas. I choose a
Coke, the last one for a long time.

I offer my gift of an ornate gilt clock to Haji Roba's young
second wife, Mumina, who has just given birth to a son. She
embraces me warmly. My daughter, Sveva, has sent two glittery
'diamond and gold' bracelets to their eldest daughter, Rehema,
that delightful graceful little woman of a little girl, perhaps ten
or eleven, who serves us with immense serious concentration,
balancing plates on one hand like a dancer.

Haji Roba offers me his gift: a tiny leather pouch stitched by
hand, which contains a page of the Holy Koran, a precious amulet to protect me on my travels; he looks me over pensively, as if
trying to work out if I will make it, on foot, through the deserts.

We finally disengage ourselves from our warm hosts, and leave
Tony at the Beaver, at the airstrip; it is rather late for his flight
back to Timau.

As a parting gift Tony hands me two cartons of sun-dried rai-
sins from his vineyards: the symbol of a European world so far
removed from the reality we have chosen. He tips his hat, chival-

rous as ever, climbs in and takes off in the dust while I hurry to change behind a malodorous, well-used outdoor loo.

Watching his aeroplane disappearing in the afternoon sun, I think that the last link with our world has now gone. We are on our own: we can only walk back over that forbidding torrid territory.

Ages before, when I was crippled in hospital, I had dreamed of just this: to be wandering in Africa, through savannah teeming with wildlife, with a romantic figure of a man – who belongs to that Africa – for my guide and companion. My dream of Africa had kept me going, yet I had no reason at all then to imagine that it would materialize.

I met this man when tragedy, upheaval and solitude prevailed; and now, in the ripe midday of my life, my physical wounds healed, but with more scars engraved in my soul, like the story of a surviving tree written with cuts and marks on its scorched bark – I am here, facing the challenge of the unknown.

I look at Aidan: his rugged face burned by many seasons of sun and rain; his noble head with its classic masculine features, his eyes of deep blue, familiar with deserts and mountains, prairies and drifting clouds; his long muscled legs and his slender figure made for effortless walking. He came into my life in days past, he went away and he has come back at my side.

That he asked me to join him in this adventure is a great honour to me: not only has he offered to share his solitude: he thinks I can make it.

Flying over the land we plan to walk through, I have some doubts: the landscapes look unforgiving, the distances immense. Will I be able to do it? Do I presume too much? Do I detect a doubt in Tony's eyes? Is the plan too ambitious? Even though my crippled leg healed long ago I am conscious that I have not regained total flexibility in my knee, and there are signs, almost twenty years after the final operation, that my leg is becoming shorter.

Whatever it is, I am taking a risk. I have now no option, and I am determined to try and live up to Aidan's expectations.

It is happening now.

I wear my clothes: new 'soft' (I thought! I was wrong) walking shoes, baggy trousers and a khaki shirt.

Our camels are waiting at Fujiyanyota, already laden by an efficient Osman, on whose head the hemostatic cotton has dried the blood, and sticks, startlingly yellow, like the feather of an exotic bird.

I admire the new camels. By the time of the full moon, Ibrahim managed to buy only eighty, not a hundred as we had hoped. Now seventy-eight remain: two escaped, when the hyena scattered them a few nights ago. Nine large impressive males, and a herd of superb looking young heifers, with dignified, graceful bearing, and startlingly white coats. I learn that this is Nature helping them to adapt to the extreme temperature. Darker coats would attract the sun, make the heat unbearable.

We leave Fujiyanyota about 5.30 p.m. Already a feeling of impending night in the reddening sky.

Ibrahim Ahmed cuts me a new camel stick from a special flexible shrub which grows at the side of the track. He de-barks it with easy strokes of his knife, and hands it over with a shy, chivalrous grin. I know that I will come to regard this stick as a friend. A camel stick to put across my shoulders, on which to hang my tiredness, clinging to it with dry hands, delivering to it all the weight of my upper body. A camel stick to love and to cherish.

We march and march, through sandy, parched country dotted with skeletal black bushes, along a deserted track.

Finally, I overtake the group of females.

I prefer walking ahead, with the feeling of choosing my own

path, rather than following behind, trying to catch up with an ever advancing herd in the dust.

An hour or so after we start one of the pack camels walking in the rear, alarmed who knows by what infinitesimal disturbance, perhaps not used to carrying a load, bolts, a strap securing his load snaps, and his burden falls off.

The females, startled, go into a frenzied trot. I see them all bearing down on me, heads high, nostrils flaring. The earth trembles under their drumming hooves. Insecure about what is happening, I just dive for the surrounding thickets, to avoid being trampled.

The load is put back, with some labour; all the bundles have been scattered.

We go on and on, still fresh; the sun sets in a blood-red sky behind the spidery shrubs, and it is night. The moon is high, and casts a white-green light on the ghostly shapes.

A pungent, not unpleasant, dry smell of camel in dry country, dung, dust, aromatic plants. Strangely, I think, like the smell of raw tobacco.

Finally, my feet start aching. My knees and joints are painful, numb, unaccustomed to walking for hours without stopping for a moment – camels do not stop.

We walk for five hours without interruption. We outspan finally before a stretch of lava (*'bule'* in Swahili), at Dololo Chaka. The men cut a *boma* of shrubs, build a fire, and we are each given a steamy mug of tea.

My legs feel like wooden posts, and the feet are agony. I fall exhausted onto my camel mattress, on which Aidan has spread a sheepskin. Looking at me with concern, he arranges a heavy blanket over me. I cover my head with a Somali shawl, drenched with aromatic sage oil to help my breathing, and sleep immediately, despite the chattering Somali.

I wake up with a pounding headache to a cold moon overhead. My watch says 1.30 a.m. I swallow two aspirins with water from my tin bottle, and manage to sleep only one hour before rising just after 4 a.m. Aidan is already up, and has gone to rouse the others.

20 September 1991
Dololo Chaka – Laga Mudama

We leave just after six. The sky is pink, the air still, heralding a hot day. It takes a long time to load the camels. The morning smells of resin and hidden things.

We walk on and on, following the track, passing a lava stretch and then a plain, then another long stretch of lava rocks. The road is dusty, with many tracks, mostly of hyena, Grant gazelle, giraffe, and many, many very large lion prints. I expect to see a lion any minute.

We outspan at 10 a.m. having walked for four hours. We have covered over twenty-two miles since 5 p.m. the day before – was it only yesterday? – in nine hours altogether.

Not a bad beginning.

Hot, dusty when we stop. So tired. Eat hot consommé, sardines straight from the tin and one orange.

Sleep a bit. Write the first part of my journal.

The place is Dololo Nure, under the hill of Kubi Maradab.

We leave again at 4.30 p.m.

Aidan rides for a time one of the pack camels.

My leg is aching.

We walk for four-and-a-half hours until 9 p.m., and outspan at a *lugga* called Laga Mudama. I eat an orange, drink almost a bottle of water and sleep immediately, totally knocked out.

Garsa, 21 September 1991
Laga Mudama – Choichuff

Wake at 5 a.m.

Slept very deeply. Headache. I guess I must get used to the heat and to marches without a pause. After my sedentary year at the desk, finishing my book, it will take a while. My leg is hurting. But I HAVE to make it. There is no going back, anyway, and I have dreamed so much of all this.

An orange. Water.

On our way at quarter to six in the morning.

Lovely overcast weather. After the first two hours I accept Aidan's suggestion and ride. Tall, tall that camel was. Comfortable, funny, undulating movement: I fall asleep twice. Wake up with a start, afraid of falling from my saddle.

Wonderful country dotted with trees. Cool and windy.

See two giraffe trotting along, Grant gazelle. Guinea fowls and francolins everywhere, but too fast for Aidan to get them.

The idea is to shoot what we want to eat until we reach the dangerous *shiftah* country where a gunshot would advertise our presence. People have been killed up here for less than a watch. Our shoes, even, could cost us our lives.

Aidan walks ahead for four hours, carrying his gun across his shoulder. He looks very tired. The flu we both suffered from has taken its toll on him too.

It rains suddenly, bliss, soaking us to the bone with tiny penetrating needles of water. Letting the warm drops run down my face and licking the rain off my hands. Soaked in a few moments. The clothes dry on me in no time, leaving me refreshed.

We outspan at 10 a.m. after four hours' march, in a lovely greenish plain with trees, called Garsa, where I can now write my diary: how I would love to have the time to just leave the track, and SEE those trees, explore at leisure, stop in their shade, discover some secret creature and magic corners. Places we shall never again have the chance to pass through, and we cannot really say we have seen them at all, but gazed at them with desire from a distance. There is a pace to poetry, creativity and contemplation, and it must be harmony of body and soul. A restful quiet. A pause.

The slaves who built pyramids wrote no poems.

But I always knew that this was going to be no easy relaxed safari. This is a feat of endurance. Am I good enough? I wonder. I would hate to let Aidan down. He trusts me and has faith in my capacity to walk on forever and to tolerate hardship. I must live up to his expectation. Damn the blisters.

Eat a heavy oxtail soup with onions made by Osman. Not the best cook this side of the equator, but he tries. Strange how hot food is welcome in the heat. Short vision of a chilled glass of white wine, the glass beaded with droplets, tinkling ice. No, we do not need it and I should not miss it.

Now, Aidan is asleep under an acacia tree.

I am enjoying this, actually. I have already grown to love and appreciate the camels.

THE PLAIN OF ARBA JAHAN, 23 SEPTEMBER 1991
Choichuff – Baji – Arba Jahan

(I did not write on 22 September; too tired.)

Off again on the afternoon of the 21st.

I walk, Aidan rides to start with. I suspect he does it to please me, so I do not feel that, when I do, I am being weak. He watches me. I know he is concerned about my bad leg.

I walk ahead grinding my teeth. Lovely country, and all the time we are accompanied by a haunting concert of whistling thorns. Eerie noise of the wind through the branches.

A few scattered Grant or Peter's gazelles; a flock of vulturine guinea fowls, necks adorned with blue feathers like lace collars, in the middle of the road track. HUGE lion spoor everywhere, hyena, honey badger, jackals and oryx tracks. An eagle-buzzard, huge, watching me from a dry tree stump; a deep blood-red sunset. Silence.

I walk ahead on the sand, glad to be here, proud of having dared.

Aidan insists that I ride and two hours later I accept.

My camel's name is Racub. A large mellow white Adjuran male. Rough saddle; hard; after the morning ride my legs are tender, my crotch painful. A strong breeze, kind to my burning face.

Curious, watching these patient, resilient camels from the top of one of them. Their strange, attentive eyes, long lashed, scrutinizing the soil they tred on with soft and gentle feet.

And miles of land unvisited and still, around and ahead.

At dusk we hit a lava patch, which goes on and on.

I have walked from 4 to 6 p.m. and end up riding until 10 p.m., for four hours.

Finally outspan just after this unending lava. We find an old *boma* in the dust, where we set our mattresses, too close to the camels snorting and chewing their cud, and to the people, talking around the fire.

A fierce moon, almost full on my face.

I do not sleep very well, but ravenously eat some delicious Italian tinned meat and pickled onions, welcome delicacy here at the equator.

On 22 September, today, we wake at 4.45 a.m.; still dark. Millions of jewelled stars. Smell of dust, camel dung, heat to come. We manage to leave at ten minutes to six, a record so far. An orange for my breakfast. Bless these oranges for as long as they last. Juicy, fresh, thirst-quenching. An orange is a luxury, a delight, a treasure to behold.

A lava stretch. Brown-black lava rocks, like burnished bronze, become almost red hot in the noon sun, burning through the soles of our shoes, a torment to the camels. Varied landscapes.

Soon, the round dark stones of lava give way to *acacia tortilis* country, soil of deep brown, with scattered limestone, startlingly white. The feeling of walking on a huge chocolate log studded with hazelnuts and almonds. Wildlife everywhere, and many tracks.

Peter's gazelles.

Two giraffe.

An oryx, like a cave painting.

Lion tracks.

Porcupine tracks, and scattered striped quills which I keep collecting and putting into my pocket, giggling all the way.

The resilience of these people.

Gedi and Mamhood, the kind Adjuran camel herders employed in Moyale, are tireless.

Gedi wears a loose, long checkered Somali *kikoi*, turquoise and blue, and a flowing white *shuka* on his shoulder. It is all he owns and he does not seem concerned. He is unconsciously attractive and naturally elegant like a camel or an antelope. The thought strikes me that nothing has changed in the look and clothes they wear, these Somali, since the time when the Three Kings from the Orient brought gifts of gold, incense and myrrh to the just-born Jesus of Nazareth.

Camel tracks, round, soft and gentle on the sand.

Having marched four and a quarter hours, I agree to ride again for three hours, and at 10 a.m. we outspan in the hottest place, Choichuff, and are instantly assaulted by ticks of all descriptions.

A very hot siesta; flies.

Ibrahim Ahmed damaged his nail, caught and crushed it while unloading, hanging red and yellow like a bloodied claw. I treat it to stop him from pulling it out there and then, the definitive solution he is ready to apply. It must be bloody painful as it is. I apply antibiotic cream, and wind a sterile stiff bandage around it to keep off the dirt. A strange sadness in Ibrahim's eyes even when he smiles.

'*Ah. Wewe doktary khabisa!*' 'Oh, you are a real doctor!' he praises me happily, the pain already seemingly overcome. They all appear impressed by my comprehensive supply of medicine and first aid.

They look good, my two Laikipia men, dressed in identical green shorts and military shirts, with their safari army gear and their guns. I notice that Lwokignei often asks Mamhood to carry his gun, but I prefer to ignore this unorthodoxy: in the middle of nowhere some rules can be relaxed, others must be kept firm.

If anything happened here, a serious wound, a terminal illness, a snake bite, there would be nothing or nobody to appeal to but our resourcefulness. No way of communicating with the rest of the world, and so far we have not met a soul.

We leave at 4.15 p.m. on September 22nd, still extremely hot. Aidan rides the first one and a half hours. My pillow is put beneath the saddle, which is now much more comfortable.

Blessed by a breeze.

I ride for three and a half hours, until 10 p.m., when we outspan on a dusty place in an area called Baji, a dry water hole.

Sitting in the dark next to the fire, I take off my shoes and, looking at the blisters on my feet and the scratches and cuts that a few days of walking wild have engraved on my body, I think once again how much I have dreamed of this during the long months and years. I massage my leg where the long scars left by several operations remain. Aidan is watching:

'It is amazing that you can walk at all. That Swiss surgeon performed an absolute miracle. What was his name? Ah yes, Professor Müller. He would be proud of you if he could see you now.' He smiles at me. 'You should write to him; thank him; send him a photograph.'

'Professor Muller: I wonder where he is now . . . over twenty years ago . . . he may be dead.'

Aidan pokes the fires, adds a log, stretches comfortably on his mattress. He crosses his hands behind his head and looks up at the sky. His dark blue eyes seem almost black. His resemblance to Paolo strikes me once again.

'Tell me about him. I like to hear your stories.'

Holding a hot mug of soup, gathering a shawl about me, under a sky bejewelled with stars, listening to jackals far away, I close my eyes. From the treasure chest where my past sleeps, a face emerges, images of places long unvisited, train journeys, hospital smells, birds pecking at my window, the impression of a smile over a small moustache, twinkling brown eyes, strong hands, my bandaged leg. Hopes, fears, pain, trust . . . and memories unfold.

Walk. Walk up and down the room.' There was a lilting authority in his strong accent, but kindness in his voice. 'No. Without crutches.'

He removed the long metal sticks I had grown used to regarding as part of my anatomy. I looked at him appealingly: without crutches I felt vulnerable, inadequate.

'Walk.'

I limped like a lame duck, hobbling on the shorter leg, biting my lip in concentration. When I turned, he was frowning. Once more he held the X-rays against the light. He had of course already studied them carefully, before agreeing to see me. Complex cases intrigued him.

I realized that – like a sculptor who dreams of the statue he will mould from a mess of clay, or an architect planning a complex restoration of a damaged building – he knew exactly what he was going to achieve, what problems he could encounter and how he would deal with them.

I looked at him with total confidence. He was my only remaining hope of ever walking normally again.

Professor Müller, the legendary orthopaedic surgeon in Bern, architect of miracles, was the best surgeon for the sort of fracture I had suffered in the tragic car crash in which Paolo's wife had died; the only one who could reverse the mess of past operations and heavy plasters that had gone wrong, who could repair my shrunken and crooked leg and make me walk again.

He looked at me intently.

'You'll need three, possibly four operations.'

He paused to see the effect his words had on me. I swallowed. After one year of plaster and two painful operations, four more operations, and the time in between, sounded like forever. Like hell. I had been crippled for over a year now. The thought of more hospitals, exercises, anaesthetics, was unbearable. I said nothing.

He looked at me, cleared his throat.

'It can be done. It will not be easy. It will take time. You must help me. You are too young to be a cripple.'

Too young. I had just turned twenty-six, Emanuele was four years old, and I felt as if I was awakening from a long sleep to the possibility of being an invalid for life.

I simply could not afford not to be able to walk. There was Paolo. There was my dream. I wanted to live in Africa. Only that mattered. If it had to take four more operations, then so be it.

I looked back at him, and I managed to smile.

'Thank you,' was all I could say.

The lights overhead blinded me. I blinked. Already, the wave of numbness induced by the pre-anaesthesia was claiming me. I looked up. The kind, intelligent face smiled down on me. Brown, mobile eyes. The surgeon's mask hid his short dark moustache. He took my hand, squeezed it. I marvelled, vaguely, at the strength of that small hand.

'You will be all right. See?' The lilting Swiss accent in his polished French. He waved a sort of measuring tape. 'Your leg is going to be exactly of the same length as the other.'

The brown eyes twinkled. A needle in my arm.

'Maintenant, contez avec moi. Jusqu'à six. Un, deux . . .' I

breathed deeply and repeated with him '... *trois, quatre, cinq* ...'
I never said six.

I woke up in my bed, a few hours of my life gone forever
from my consciousness. My leg felt heavy, loaded but not really
painful.

My mother, who had come from Italy to help me, looked at
me with concern.

'A much longer operation than he thought, he had to combine
two in one. Four hours.'

The door of my room opened gently and he came in. Still
dressed in his operating outfit, he marched to my bed. After the
formal pomposity of the Italian dons, who arrived escorted by
bowing assistants and a retinue of nurses, his simplicity took me
by surprise.

'*Ma chère amie* . . . *ça va?* Stand up and show me how you
walk.'

Incredulity, uncertainty, fear of falling. Standing and walking
immediately after such an operation seemed impossible. He
helped me up, fixed the drain from my wound to the belt of my
hospital gown. He held my hand. I stood.

On the grey linoleum floor of that white-and-steel hospital
room of the Lindenhof, with a dancing heart one afternoon of
late July, I moved the first tentative step of my new life, no longer
a cripple, the first step on my way to Africa. It would, one day,
be the grassy plains of the savannah, the coral barrier of the
Indian Ocean, the Great Rift Valley slopes, and now, the sand of
remote deserts, the lava rocks of the North Frontier tracks. I owed
much of my life to him. I blessed Professor Muller when I had to
run fast from a wild beast, to climb a tree.

Professor Muller. I smile at Aidan. An extraordinary man, an
inventor and a wizard. I wrote about him in my first book, just
to thank him; I doubt if he ever read it. I wonder if he still works,

if he is still alive. He seemed to me old then. I don't think I will ever see him again. It was all so long ago.

I lie down on my camel mattress, covered with a bed sheet and blanket. The sheet is a luxury which makes all the difference to my aching limbs.

The past is past, and this is my reality.

Far from the noise of the camels and the *boma* – thank you, Aidan – I spend the best night so far.

Wake up very early.

We manage to leave by 5.30, a record. Hot and very dusty track. I am determined to walk, the camels make me lazy.

Yesterday I made up many stories for my next book; I find this helps to distract me from the tiredness and weariness of walking in near silence for hours. Yesterday I imagined 'The Bull Shark of Vuma' and 'Fifty Guineas' Pike', and completed 'The Full Moon Island'. Would they ever be written? Am I creating a new book, here in these desert sands?

Today I did 'The Story of Nungu Nungu'; as Aidan likes it and suggested I should write it, I shall dedicate it to him. I completed 'Emanuele's Chameleons', and I began in my head 'Only Dust at Arba Jahan'. This last one is prompted by the ever pervading dust. It could be the title of this safari's chronicle.

But how gently the camels walk. If they were cattle the cloud of dust from their hooves would be totally unbearable.

Long, long march, as we finally turn due south-west, on the long, unending desert track after Arba Jahan. We have to be careful to avoid the wells. *Shiftah* will be lurking there.

I admire Aidan's proficiency with his old brass compass, how he takes our bearings: it seems impossible that we shall ever get out of here.

We meet today the first man since we left, four days only – but they feel like four months, time has lost its meaning – a friendly camel herder of the Adjuran, sturdy-looking, carrying a curious leaf-pointed spear. He wants to sell us a camel, even on credit.

We outspan at the Arba Jahan plains, a windswept open area dotted with rare trees. Grassy straw. All the camels decide to come and feed from the tree we have chosen for our noon rest, one of two only, and far from each other, in a vast waterless plain. I am now writing here, having changed to a fresh blue caftan, closely observed by ruminating camels.

A blessed breeze: otherwise the heat would be unbearable. A few herds of camels file away in the heat waves and the distance, like mirages, silhouettes in a biblical scene; they must see us as we see them, but they make no move to approach us. They must be heading for the water hole at Arba Jahan, over ten miles due east, which we avoid as we have heard that it may well be dry, and full of desperate livestock, and starving Somali.

They will find only dust, at Arba Jahan.

Our valuable camels and our daring plan must be kept to ourselves. We do not want to advertise our presence. Ahead is *shiftah* territory.

This tree, a Salvadorensis, underneath which I am writing, is possibly one of the last trees for days. At this time, if he is not too exhausted, Aidan usually takes our bearings with his instruments and compass.

It is an old brass compass which I gave him, found in an antique shop in Nairobi, and which surely has a story of explorations in wild lands. I wonder where it has been and what were the eyes of the man who owned it like? What did he think, how did his voice sound, and was his woman fair?

Now, in the far distance, a ghostly evanescent mountain shape may be Marsabit. Aidan shows me a gaping void in the map, marked with lava symbols, a desert we must cross. I can see he is worried, he asks our people to fetch and load *kuni* on the camels, so that we can light a fire in the day – days? – to come, for our tea and their evening *posho*.

No trees mean no firewood ahead.

Dust, heat, thorns, a seemingly unending old livestock track, and finally the desert, lie beyond.

Just enough water to the wells of Koiya?

24 SEPTEMBER 1991
Arba Jahan

We started walking at 4.30 p.m., yesterday afternoon, leaving behind the last yellow-grassed plains of Arba Jahan. Arba Jahan means 'Three Elephants', but there are no elephants left in the arid and cracked soil of this unwelcoming landscape. We have promised Tony Dyer to note any sign of elephant in our route, but so far we have seen none.

I like this place, however, its space, its silence, the feeling of walking on high land.

I am glad that we changed route. I feel like holding my head up and sniffing deeply, sensing the breeze, as the camels do.

Dark and dusty terrain, cracks covered in straw, full of the white, conical shells of ground snails, which crawl in vast numbers out of their holes in the rains. Thousands of squirming, slimy snails. Revolting thought. I have always loathed snails.

We walk for two hours on the plains, where Aidan sees a cheetah with two cubs in the distance through his field glasses. But when I look, it has vanished. Mirages are part of the landscape here and encounters are glimpses, mysterious and evanescent, they escape long scrutiny.

I walk rather than ride, as I had in the morning, and feel in great shape.

At 6 p.m. Aidan shoots a Peter's gazelle, a young male, for the *watu*'s meal. A last treat before the desert.

It happens so fast. We spot a small group of gazelles in the distance, watching us with high necks, as still as statues. Without pausing a moment Aidan nods to me and gives me his hat to keep so that we can get on walking without changing pace, not to alarm the gazelles. There are about eight in a small herd, perhaps half a mile away. He pulls out his .22 without altering the rhythm of his step, quickly and precisely, with fluid movements, as a hunter must. We hear just one shot, quick as a whiplash. Then I see Osman leap high like a grasshopper and dash through the plains with a young hare's sprint towards the creature, to cut its throat as Muslim custom demands. Without this *halal** practice they cannot eat it.

Young Gedi, one of the two junior camel herders, runs after him with incredible speed, *shuka* flapping in the wind.

The horizon seems close and far away at the same time, and there they are profiled on it. I move towards them across the immensity of the plain.

The gazelle dies instantly, hit through the heart. Just enough time for Osman to slit its throat with Aidan's knife. They cut off its head, and in the dusty pupils, opaque and open in the twilight dusk, I can detect no pain. The memory of other dead dark eyes in the sun.

Jubilation, as meat equals a feast to all Africans. They throw the dead gazelle across the saddle of one of the pack camels and we move on.

We march and march for five hours through lava, and finally outspan at 9.45 p.m. in the middle of the track, the only possible place free of rocks.

A full moon, whining jackals and wild dogs and hyenas filling the night with prayers to the sky.

A lion roars at 12.30 a.m.

Wake up at 4.45 with an orange to wash my mouth of sleep, and leave before sunrise.

*Slaughtered according to the Koran rules.

When the sun rises we watch it.

The desert lies ahead inexorable, forbidding, simmering in the distance. We cannot see its end. The sky is salmon pink. We stop a few moments to consider the challenge before beginning our march through it. Aidan turns up to me, our eyes meet and hold, and I nod. Just that.

'*Oho, galla,*' he prompts them. The camels snort and we move on.

I only walk for two and a half hours, my feet are aching so much after last night's five-hour march and the five hours in the previous morning. I do not think I could have continued to walk this time. The heat is scorching. The lava rocks burn through the soles of my shoes like live coals; Aidan's shirt sticks to his back, its green colour already faded and discoloured.

I agree to ride. All the men are silent.

My camel's name is Racub; quiet fellow. I ride, with drizzling dreams, falling asleep in the saddle, for three and a half hours. Aidan walks, showing no sign of tiredness. This desert seems to stretch over the horizon forever. The swelter lifting from the stones at noon is extraordinary. Heat waves like mirages of water tremble on the horizon, trees appear in the distance, dark, with promise of shade, but we can never reach them. They are illusions of the fever of the heat, substance of dreams or hallucinations.

The sides of the track are burned by some exhausted fire; ashes and carbonized tree stumps add to the inferno feeling. Surprisingly, strange oblong nests of invisible birds festoon the majority of the skeletal shrubs like hope, transforming them to ghostly Christmas trees.

Where are the birds, I wonder, and what can they eat or drink? Not even insects can survive in this heat.

In my head I write 'The Secret Cove', a story of ocean, water and of long-lost days.

Driven by the fear of being held in that hell, we make it much faster than we thought: time loses any meaning when walking through endless, monotonous stretches of desert like this one.

We outspan finally in a god-forsaken lava stretch, after crossing another desert of burned grass and whistling thorns. The sun is high on the horizon, unforgiving.

The camels stand silently next to us, with bowed heads, without eating, searching for a sparse shadow, and elusive moisture. We make an illusory shelter, tying my green canvas *tandarua* to a low dried-out shrub and to the handles of our food boxes. We slide under it, no more than a foot or so, on the stained camel mattresses, and Aidan is suddenly asleep.

25 SEPTEMBER 1991

I walk the first two hours, having left the desolate lava stretch at 4.20 p.m. I am getting used to walking.

The camels take a very fast trot, and the men sing and whistle in high spirits, galloping along as fast as them, and probing them now and then with their camel sticks.

Although I feel really well, I just cannot keep up and finally have to ride. It is a painful ride, as I am sore and hot.

Beautiful country, though, open, wild, with yellow grass.

Is the rest of the world still there?

Outspan at 10.30 p.m. on a pleasant enough open area, with tall dry grass.

The camels begin to suffer from thirst and too much heat. They are splendid creatures, and I am full of admiration and respect.

We are treated to the classic dish of liver, kidney from the gazelle and onions, cooked by Osman: but it is inedible, too salty, strong and gamy. A powerful after-taste of urea.

The men sing and laugh late into the night around their fire.

Outspan in the heat of noon below an acacia tree facing silent hills, live with wandering insects.

Leave this morning, 25 September 1991, at 5.10 a.m., having woken at 4 a.m. The camels doze nearby, nibbling from sparse shrubs in the stillness.

Aidan looks preoccupied while consulting his maps and searching the horizon for landmarks. A far chain of unknown hills is all we see. Where is water? Will we make it to it? Will the camels survive?

We have apparently taken the wrong track, and will have to cut across country, a thought I relish, as this will free us from the monotony of following a marked line, however faint.

But what if we get lost?

Aidan finally sights the lava ridge which is our landmark to Koiya and water.

We should reach it in two days, but I do not dare to hope it.

26 SEPTEMBER 1991

Walk through much nicer land, seasonal swamps, not easy for camels as the dry shiny grey swamp-grass is slippery and hides some dangerous holes.

Aidan knows we have missed our track and is anxious about losing precious time, concerned about the difficult terrain ahead. Tall tangles of thorns and spiky shrubs reach the camels' shoulders and are practically impenetrable, dangerous as they could hide deep cavities and the camels could break a leg. Vicious little hooks dig scratches through our torn shirts.

But we must go ahead without delay. Water is the key. What will we do if we cannot find the wells? Aidan's mouth is set in a grim line. But I am in much better spirits, as the landscape is so lovely and African and the pace much slower so I can easily walk. I secretly rejoice at our involuntary mistake.

Suddenly, on a straight stretch amidst tall acacia and thick silvery swamp grass, in a glade of afternoon light there stands a magnificent animal, unafraid and graceful, a male gerenuk. A fairy-elf of the savannah, who can only bring us luck. We watch each other for a while, then he leaves at his own pace, in elegant, unhurried leaps.

An unforgettable sight.

We leave the main track to cut across country. Luckily the landscape has changed, open now, with no more twisted shaggy bushes to impede the camels' progress: thanks, Emanuele! I always appeal to him when I am in trouble. Aidan has been so worried that I have to cheer him up. I just know it will be all right. The camels patient, wise, gracefully bear their burden in long strides. My optimism is rewarded: a good lesson.

And then, in answer to my silent prayers and to make Aidan happy, we come to red sandy soil, dotted with acacia and comiphera. Totally wonderful walking through this terrain, resilient and solid soil allowing a bouncing step. My spirits lift as they have not in days. We outspan on a vast *mbogani* and light our own fire for the first time, while the men cut the usual *boma* of thorny bushes which will confine the camels for the night. Darkness falls on my happiness at having left the dreary straight line. The moon rises white and silent.

Then Aidan grimly announces that thirty camels are lost, and the herders are following them. They kept on by the light of the moon when we stopped, aiming towards far hills, which to them perhaps meant water. Like ghosts in the silvery light, they moved on, drawn by inexplicable instinct. There is nothing we can do but hope that our skilled people will catch up with them in time.

I try to cheer Aidan. If the herders can find the tracks by moonlight, they will follow them and we will wait here for a day. Not a bad thought, relaxing in this lovely country. Aidan is pessimistic. I reason that we are far from water holes and the confusing footprints of other camels, and that it will be easier for our skilled herders to follow the pale silvery tracks in the moon-light. But if the camels decide to go at a brisk trot, looking for water, it will be hard to catch up with them. They could be found by *shiftah* and stolen easily. It has happened before. And if the large lion whose footprints we detected on the track decides to ambush them, as is quite likely, they will gallop off in terror and we shall never see them again.

We go to bed early, having spilled the water, the salt, the muesli and the pasta!

But we underestimated our people; at about 10.45 p.m. we hear calls, whistles and screams. The night is full of activity. And we know they have found the camels.

We leave at sunrise on the 26th, happy, carefree, in excellent mood, through attractive pale yellow grass plains, rusty sandy soil. Compact and easy to walk on. Reach the lava boulders.

Outspan in a sandy *lugga*, under a clump of *Salvadoria* trees, a delightful refuge, and I feel happy. Ahmed offers me a special twig he has cut out of the tree's fibrous root, a valued original toothbrush to chew on. I am happy. Alive.

Aidan has learned the value of drinking water liberally during this safari, and although he is not at all well, he certainly has improved enormously. But he has blisters on his back, legs and his poor *sedere*. Bad suppurating tropical sores from scratches gone septic, which make walking painful with the chafing and the heat. I wash them with half a cup of warm water and apply antibiotic cream from my kit. I have convinced him to wear a loose *kikoi* rather than trousers, as our Somali do. He looks great in it, long-limbed as he is; and I feel in superb shape.

MATOKOLE, 27 SEPTEMBER 1991
Still twelve miles to Koiya

Last evening having outspanned on a plain before dark, Aidan shot a francolin which we devoured, grilled on an open fire, impaled on a rough wooden spit and spread with salt. A feast.

At bedtime, as I lower my head onto my mattress on the ground, I glimpse a quick movement, like a flutter, by my face. I have learned, after years of Africa, that instant reaction can save your life. Without thinking, I instantly sit up: a large pale yellow scorpion, the lethal *Sole Fugens*, disturbed by my gesture, his poison-laden tail arched, runs out of his stone, straight into the fire, where he sizzles to death in a few seconds.

We walk through the most beautiful country this morning: no sign of livestock or men because there is no water. But full of game. Grant – or Peter's – gazelles, oryx, giraffe, a landscape of tall fantastic acacia, open red-gold sands, a divine place. Walk for seven miles, between two lava ridges, of outstanding stark beauty. Must be wet seasonally, as there are signs of old livestock and MANY lion tracks.

Aidan tries to shoot some guinea fowl and francolins, but they are far too clever and fly off.

We take some photographs of the camels moving through acacia; then Aidan discovers that his camera has been empty all along! Tough luck. I have some good shots in mine, I hope. Now outspan on a sandy *lugga* of large acacia not far from Koiya. A disturbing feeling of people around. Although they are invisible, it seems that we can hear livestock bells.

The camels are all around me, lying in a circle and observing all we do. They nudge my head. They placidly guard me while I write. I wonder if they know how close water is: but is it?

Koiya, 28 September 1991
KOIYA!!!!!!!!!!!

Water at last, and TWO days ahead of schedule.

Clean, fresh, abundant, sweet water at Koiya.

We marched yesterday evening along a lovely, grand sandy *lugga*, Matokole: difficult walking, and I refined the skill of finding the hard bits of sand – usually a darker grey crust – on which to walk without sinking.

We meet four youngsters from the tribe of Rendille in bright beaded ornaments and red *shukas* – the first people, after that Adjuran herdsman. Then, their scrawny white cattle, in clouds of dust.

Lion tracks – these omnipresent but invisible lions – gerenuk, but incredibly attractive banks along the river, with trees and palms, huge acacia and enamel-green bushes. I wish we could stop below them, but we have to go on.

Instead, we spend the night on a small rise of white pebbles and grey soil; Aidan shoots two francolins in the *lugga* – and two for the men, who rush to cut their skinny throats – and we roast them on a basic spit on an open fire, a royal meal, plus half an orange each and hot cocoa.

Wake up at 11 p.m. with a fat tick stuck on my shoulder, gorged with my blood, little swine; amazingly, the first tick to bite me, in this tick-infested country, and we always sleeping on the camel mattress thrown on the bare soil.

Sleep soundly thereafter until 4.30.

Leave at 5.45 a.m. on the 28th September.

The camel bells tinkling away like birds in the noonday sun, we proceed, following countless livestock trails, all converging on this famous place, the only water in miles and miles, but the herds are always ahead and never to be met.

At 7 a.m. we come across a Rendille *boma*, tiny, with goats still crouching. A sleepy, cheeky young Rendille wrapped in a red *shuka* is milking them with lazy strokes, the sun already high, to Aidan's indignation. He cheerfully greets us with no curiosity as to how the hell we could materialize, two Europeans and eight Africans carrying guns, with our seventy-eight camels, in the middle of nowhere.

Koiya appears at 8.15.

We stand for a time in silence, overwhelmed, to watch the target of our unforgiving progress.

A far outpost of the government. A group of *mabati* rondavels scattered on a rise, goats, a tiny *duka*, some round huts from whose dark doors people watch in silence. Luscious doum-palms on a *lugga* and clumps of tall acacia trees advertise the wells, like the oasis in the Sahara I remember visiting once with my father. That is it. We have made it to Koiya, after all.

From clumps of rounded dwellings that look like the yellow cocoons of giant insects, a thin figure in long robes comes to meet us, holding a ceremonial stick. He is a kind old fellow called Hassan, the councillor who – guess what! – knows our friends Jasper Evans and Maurizio Dioli, and buys camels for Jasper.

In Muslim tradition, he greets us with deference as weary travellers, and accords us the courtesy our status demands. In no time – and with some fuss – he organizes his well for us, and a secluded patch of trees to outspan under.

The well has been dug in the sand, the opening lined with round borders of a pale compacted clay, and the water is transparent and yellow like a topaz, incredibly inviting and, for me, forbidden.

The camels drink in disciplined shifts of four at a time, waiting patiently for their turn in the shade of the acacia, while we watch and the two Ahmeds sing the haunting Song of Water. This is an antique hypnotic tune of celebration, evocative and melodious, which is sung whenever camels are approaching water: for the camels it means drinking, and, obedient, they perform accordingly. If camels could laugh, the air would resonate with their nasal whines of jubilation.

I really admire the control and dignity of these creatures, thirsty as they are, who let their companions drink their fill before they move forward. They curve their gracious white necks and drink rhythmically, unhurriedly from the troughs carved out of hollowed tree trunks. Their caved-in stomachs inflate gradually before our eyes; the bony ribs disappear, and those haggard tired beasts are transformed once more into handsome, fit young camels ready to start the endless march again. They begin to feed instantly, peeling small leaves off branches with spongy lips.

Aidan buys our people a goat, assisted by Hassan who does the bargaining, and we shall feast on it tonight. Osman is busy frying oil in a pan.

Liver and kidneys and onions and rice will once again be our spartan/glorious meal.

Aidan goes up to his waist into the yellow tempting water, helping to fill the troughs, and I envy the men's freedom in the Muslim world which does not allow women to bathe at the wells, for fear of defiling them.

Now: the first bath in eight days any second, any amount of water, and I shall wash the dust off my hair, behind a charming screen Aidan is building for me, a perfect retreat, sheltered from people's eyes. Now and again he smiles at me. His beard has grown and gives him the patriarchal look of some biblical prophet.

I love him.

TORNGONG, 29 SEPTEMBER 1991
Koiya – Torngong

We spend a blissful entire day and night at Koiya.

We set up our little den by a clump of trees, under a mosquito net, the first private and comfortable bed since we left. The men sing low and hoarse songs of love – of war? – around their fire.

Earlier I noticed, in an abandoned seasonal *manyatta*, curious contraptions of tall scaffolding made from branches; Aidan explains that they are suspended beds, to protect people from mosquito, which do not fly that high.

So I suspect the night will be disturbed, but in the event, not a single mosquito is heard. Too dry at this time of the year for them.

Soon after settling down and arranging our camping gear, the Rendille come with their amazingly skinny goats, but they soon pass on, discreet and unbothered by our presence, and Koiya is all for us.

I observe these goats and the most rachitic looking sheep, nibbling at invisible morsels of food, never lifting their patient, bent heads from the dust: they seem to eat just that dust, but Aidan informs me that in fact they feed on fallen acacia flowers, so small and inconspicuous as to be like powder. Extraordinary survival skills.

A girl comes over and I give her a sweet. She takes it shyly, runs away, and soon three more girls are there, waiting in silence. I give them a handful of peppermints each, and they go off together, giggling like tinkling bells.

I sleep, write, take it easy. Wash my hair and all my body, luxuriously, lazily and happily, with plenty of water. Change into a clean caftan; walk down to the wells to see the camels drink, and generally regain my human condition.

Leave at 5.45 a.m., having woken up at 4.30 a.m. after a disturbed night as the camels – too close – never stop thumping, and the men have forgotten to muffle the wooden camel bells with handfuls of hay.

Walk at a steady pace to the lovely hill of Sepi, boulders of quartz with trees in the smouldering heat.

I pour water, now fairly abundant in our jerricans, all over myself and my shirt and trousers, and I feel renewed. I am dry in a few minutes.

Soon on our way to reach . . . Kairu?

Now, on the evening of 30 September, sitting out on a plateau overlooking the hills. Hills like a stage, empty of actors. Close to

an old *manyatta*. I write in the last light before dark, while the
men are cutting a *boma* to enclose the camels. Lions are about
and we could have a stampede.

30 SEPTEMBER 1991
Torngong – Lontopi

Lions roared continuously last night at Torngong, very close to
us, ready to attack; the night resonated with their hunting growls.
The camels stayed awake, nervous in the dark, ready to stand and
scatter in all directions. We had to build a stronger *boma* and the
men slept in turns, with fires all around. I lay staring at the stars
overhead, sensing danger, and sleep fled.

In the morning their pug tracks mark the dust around the
boma. We leave at 5.50 a.m., and walk only three hours and a
quarter. It is on this track that we find the first unmistakable
footprints of *shiftah*: three people walking alone, with no livestock, wearing the characteristic rubber sandals, fashioned from
old tyres, that they are known to favour. The men become quiet
and wary. Aidan loads his rifle. We all keep looking around,
uneasy at the silent, enigmatic hills approaching Lontopi.

Despite our concern, when we stop Aidan builds a shelter for
me on the side of a reddish *lugga*, and I wash again with great
relish while he keeps guard. My hair dries in minutes.

LONTOPI, 1 OCTOBER 1991

This is the idyllic place called Lontopi, in Samburu land where
there are ponds among the rocks; where there are springs and
palms and shady trees along the hills; where in the sun one can
wash, and let emotion prevail.

On our way to Kairu, back on schedule now: not many miles

– just over twenty – covered in the last two days, as the terrain is uneven and rocky and lion tracks are everywhere.

We shall now have to work hard, walk more, to keep up with our plan. But it was worth this rest and indulgence from 9 a.m., and the memory which will not fade.

These large camels, I observe, love to feed on the most improbable grey, desiccated weeds – *Indigophera Spinosa* – which must be powerfully nutritious, but look like cobwebs. I watch them nibbling at these frail delicate grasses with their prehensile lips, chewing them again and again patiently for hours to extract any nutriment they may have.

They munch away in great content all morning, and drink at the springs at Lontopi, while I watch them, perched on a rock like a baboon.

A beautiful mountain ahead, with great boulders and some doum-palms in the vast pink sand *lugga*.

Shiftah hide on these hills to survey the land around for their quarry. We could be that quarry, in fact. An eerie feeling, possibly being observed by cruel, invisible eyes, narrowed below lowered turbans. Images of cowboys in a Western movie being ambushed by impassive Apaches sitting on ponies from hilltops.

Everyone is alert and the men hold their guns carefully.

I try not to think of their AK47s and the long curved knives they use to cut their prisoners' throats. Osman, his eyes like slits, comes and talks to Aidan in hushed tones. They turn to look at me and I understand that I am the topic of their conversation.

The night is throbbing with crickets and strange unknown insects singing their diverse songs of life. I write by the light of my little diary cover, fitted out with battery and bulb, bought in a camping store one May morning in Washington, D.C., where I wore an Armani suit and had an appointment with a senator: days impossibly far removed from this reality. I notice that Osman has placed himself closer to us on my other side. Aidan is next to me, his loaded rifle at hand.

The night vibrates with tension.

I wake up early, stiff and covered in dew. I see Osman's eyes on me, in the livid light before dawn, and know he has not slept.

I sit to finish my diary, while he revives the fire for tea.

For as long as I live I shall remember images of this unusual and unrepeatable safari – which has gone almost, already, in a soft haze of days. Places unvisited by man, reachable only on foot. Far from roads, miles and miles behind hills and *luggas* and waterless plains; lava hills and quartz sands, stones glittering with green and coral fire.

Patient camels, resilient, trustworthy, wise with the timeless instinct of survival in the harsh forbidding lands where they belong.

I shall remember oryx trotting off up hills, black tails flickering; the gerenuk surprised in the swamp, in the narrow path of grey silvery cane-grasses. The Song of Water at Koiya, exultant, melodious, and the jutting, darting walk of the wild Rendille warriors clad in red, with their fine faces and lithe bodies, muscular elves of the sand *luggas;* the starving goats and sheep searching for invisible particles of food amongst their own pellets; the gentle rhythm of the camel bells; the methodical unhurried chewing of their cud; their eyes, fringed with long wise eyelashes, their way of kneeling on the dust, sudden, complying. The funny shape of their sturdy tails, like huge silverfish; the way they urinate, backward, letting their brown gummy urine with its aromatic penetrating smell of unknown herbs trickle down their back legs to evaporate and cool them; the oily sap of the spicy *ginau,* the balsamic plant tasting of mango and turpentine; the colour of the rising sun in skies of red, the silhouettes of the camels against the horizon; the morning light on the pale yellow grass streaked with silver; the dust and twigs in the eye of the dead Peter's gazelle; a little natural shelter on the Arba Jahan plain, below a *Salvadorensis* tree, surrounded by the camels browsing on its leaves. The quest for Koiya and the elation, the sense of achievement, at reaching it ahead of time; the colour of the first water in the well, topaz yellow and transparent, so good to drink and fresh on my sunburned skin.

The Rendille girls, heads small, circled by bead ornaments, rusty red peplums over brown leather beaded skirts.

Our fires in the silence of the night.

The emotion of the pond at Lontopi.

The long ride on the endless desert.

Walking, step by step, in Aidan's tracks, the pattern of the soles of his shoes, familiar round marks on brown rubber.

The lion's roar with the moon. Its pug marks printed in the dust. Its feral smell brought by the breeze.

The ever present fear of invisible *shiftah*. The taste of adrenaline when we see their tracks.

Osman's eyes, coloured by his courage.

Tonight I am writing lying on the sheepskin on the camel mattress; the fire smoulders. A quiet intermittent touch of camel bells; songs of nocturnal insects. The sound of my breathing.

A breeze, like waves through the palm trees. The flavour of wonder.

Aidan, asleep at my side.

2 OCTOBER 1991
Lontopi – Il baa Okut

At Il baa Okut, our second to last night in the bush.

Walk through Samburu land, across the Kairu *lugga:* huge expanse of white sand, doum-palms in large clusters and an almost dry well circled with tangles of thorns.

We meet two Samburu women and one donkey going for water. A biblical scene. Then a little boy in a red *shuka* herding surprisingly fat calves.

Hot, hot and very aching feet. Amazing views of the Ndoto mountains and Lolokwe in the distance, our target tomorrow, and the end of our journey, where Karanja will be waiting.

3 OCTOBER 1991
Il baa Okut – Lolokwe

We walk along the pink quartz sand *lugga* of Il baa Okut, extraordinary old trees with hanging roots, on the banks amongst tall hills made of red granite boulders, emerging from a sea of green bushes. Osman walks ahead, leading Racub, a youthful spring in his step. He turns to smile at me with true liking, and I am touched. I can well see why Aidan, who has been here before, so much loves this place.

The surprise of a lesser kudu, motionless, watching us from the middle of the *lugga;* a grey monkey, darting agile across, and up a tree.

Tracks of cattle ahead, but we never seem to reach them.

I have grown to dislike cattle during this safari. Their total unsuitability to this land, and what they do to it: a lifeless, desolate landscape of dust and dried dung, flies, ticks, hoof marks and torn shrubs. The camels on the other hand, majestic and solemn, sail on disdainfully, leaving only their soft print, which does not raise dust, docile camels who know their names and people's voices, noble creatures. I am totally sold on camels this trip.

Now, waiting under a tree just before the rendezvous, where my driver Karanja should be already, with water, milk, *posho*, vegetables . . . but is not, so we send the two Ibrahims ahead to guide him here.

Aidan builds yet another bathroom for me, the last, the most charming of all, with my green screen and stones for the shower's floor, and I wash my hair, soap in bubbles, and I feel fresh, civilized. I wear a red caftan for happiness.

The camels are grazing quietly nearby, watched by those tireless herders, Gedi and Mamhood. Osman silently brings me a mug of sweet strong tea with ginger. We wait for the car from Laikipia.

We have today, at 10.45 a.m. completed 280 miles in thirteen days. Twenty miles a day average, though in the first six days of the safari we averaged twenty-seven miles a day; we slowed down when we stopped walking at night and left the track. It seems impossible to believe, but I actually made it.

The noise of an engine.

Karanja has arrived with my Toyota.

And this, then, is the end of the safari.

Later we send Karanja with Aidan's Samburu guide, and Ibrahim, to buy a goat in a nearby *manyatta* of Samburu Moran.

Tonight a last celebration. Then, Laikipia. The camels will proceed to Aidan's ranch with the staff: the remaining itinerary is not so intriguing, because it is not new to us, and we shall go ahead with the car. Karanja says there has been trouble with the government, and possibly Saturday there might be more unrest. It seems so far away and pointless.

4 October 1991

We feasted last night on the goat, with rice and fresh cabbage, and fresh milk with which I made Aidan a surprise treat of a chocolate mousse, having smuggled a packet in my saddle bag and saved it for the last meal.

Extraordinary how just a few hours of rest make one regain one's human condition, and the pleasure of appreciating the gift of wilderness. How one looks at things and savours them again. Our people have been great. Their prompt smiles, their hot teas, their morning songs.

Osman, a born leader, knowledgeable and wise, loyal to the end. Gedi and Mamhood, the two Ahmeds, Ibrahim and Lwokignei, my good Laikipia men, to whom I must give new watches as souvenirs of the trek. I do not think they have ever owned one.

Finished now. Yes, I am sorry it has ended so soon.

I only wish there had been more time, at least after Koiya, just to look around, to pause and perceive the magic of our surroundings. But the camels do not stop. One has to follow, otherwise one is quickly left behind. And camels are of course the only means of transport through desert lands, and the only chance to arrive here. So, there is no time to ponder on a bird or a tree, the gracious curve of a bank, a sudden view of hills and the shape of those oryx in the distance. An unhurried hurry is essential, inexorable and leaving no quarter. No chance to treasure the fleeing moment in the 'here and now', as the 'here and now' goes too fast, like a speeded-up film.

Like a marvellous book of complex and fine drawings, each leaf of which would be worth the pause of the attention, but the pages flick by too rapidly, without allowing the time to be observed, absorbed and known.

Were it not for this scanty diary, scribbled in exhaustion in noons too hot for wisdom, in nights of distant jackals at the light of fading embers, perhaps I would not remember all the pauses, the views, some fugitive landscape already left behind by our unforgiving progress.

Yet I shall remember what I have not written, too, as so complex and resourceful is the mind. And the colours and sounds and lights will come back unexpectedly years from now. I shall value these memories, and I know that when I am old and grey I shall look at my camel stick, shiny with wear, with deep nostalgia. And wherever I am, I will smell the resin and dust of the plains at Arba Jahan.

I am waiting for a welcome sound, the tinkling of his camel bell.

ISOBEL BURTON, letter to Lady Paget

I turn the last page. There are other, more recent pages written in ink.

LAIKIPIA, 30 NOVEMBER 1991

He survived the camel trek through lava deserts and we managed to avoid *shiftah* at the water holes, all the way up to Il baa Okut. We never met the bandits. But they caught him when he went home on leave, to rest. It was the time of the tribal clashes between the Dogadia and the Adjuran clans.

Shiftah ambushed Ibrahim Ahmed on his way to his *boma*, not so far from Moyale, and they slit his throat. I never saw him again.

They stole his new watch.

LAIKIPIA, 24 DECEMBER 1991

Aidan flew low over the house this morning and landed at Kuti before lunch.

'Can we take the car?' He smiled. 'There is something I would like you to see. Over the Corner Dam. I'll drive with you. Bring Sveva.'

We drive off in the late morning sun, leaving my house where the aroma of roast turkey and tortellini announces Christmas dinner. Even after twenty-five years, I never got used to a hot sunny Christmas, whose traditional tinsel decorations, made for grey snowy afternoons, look so totally out of place.

We meet a herd of impala and some zebra and eland and elephant coming to water. But on reaching the last stretch of carissa shrubs I see them, and am totally caught unaware.

A grinning Osman, leaning on his camel stick; Gedi and Ahmed Nyeusi, smiling, loosened turbans on the proud heads held high; and behind them, the camels.

Tall graceful Adjuran female camels, led by Racub.

With a lump in my throat I turn to Aidan. I always thought those camels were for him.

'I had no idea . . .'

He is watching me. His serious, intent eyes hold a smile:

'The breeding females were always meant for you. And Racub. Sorry I can't wrap your present. Happy Christmas.'

He hands me an envelope. There is a card. A photo of myself at the wells at Koiya, with the camels drinking in the background in that blinding white light.

On the other side is written:

'Christmas 1991. For my love, fifty camels.'

Postscript

London, 15 March 1998

Osman, The End

One evening, in early February 1998, Osman Nguyu Dupa flew into Ol ari Nyiro with Aidan. He spent the night with Issak Ngolicha, our cattle headman, to whose clan he belonged and whose friend he was.

I saw him next morning at Kuti.

He had come to greet me. He carried a camel stick shiny and noble with wear. His traditional dress made him look taller: a long Somali *kikoi,* a shirt, a dark jacket, open sandals. A white turban of light cotton muslin was knotted a few times loosely round his head. I noticed that the hair showing beneath it was black, as if the grey I recalled had disappeared somehow, and he looked younger than I remembered him.

He was uncharacteristically cheerful, more talkative than ever before. He asked after Sveva, about what I was doing. I told him that I was completing a book – this book – and that the story of our camel trek of 1991 would be part of it.

He seemed extremely pleased.

He took my hand in his and shook it a few times vigorously:

'*Asante sana. Kukumbuka ni mzuri. Asante sana.*' 'Thank you, it is a good thing to remember.'

He repeated this again with his deep, solemn voice. His serious, sorrowful eyes shone with a rare smile which spread over his ancient face with such luminous intensity that its memory still haunts me.

He was going on leave, to Moyale, for two months. Things were bad there. There had been lots of *shiftah* incidents in recent months. Many people had been murdered and a killer strain of malaria was rampant after the El Niño rains. All the same, he was looking forward to seeing his children. His teeth were startlingly white when he smiled.

It was the last time I saw Osman.

The news came to me from Aidan, one day in early March 1998, after sunset. He told me on the internal radio network, on the eve of my departure for London to visit my daughter and bring this manuscript.

Osman had been taken by *shiftah* and was believed dead.

The communications with Moyale were interrupted and telephones did not work. Because of the floods all roads were cut off. It was over ten days before one of my staff returned from leave with accurate news.

The Ethiopian clan of the Garreh has for some time been in the pay of the Tigre Government of Ethiopia, in order to report on the rebel Oromo clan, a related tribe, Borana-connected and speaking the same dialect. The Oromo rebels have been retaliating. Cattle and camel raids are the order of the day. People are killed. The Oromo *shiftah* have guns and knives and they know how to use them.

One morning, during Osman's leave, in full daylight, the sky seemed to explode, detonating with screams and shots and cattle's and camels' cries. The Oromo *shiftah* were attacking the *manyatta* where Osman kept his livestock.

He was a few miles away and heard the commotion. Immediately he realized what was happening; he ran to help. They tried to restrain him and failed. With him went a woman whose son, for whose safety she was concerned, was staying at the *boma*.

They were on the road to the *manyatta* when they were ambushed. The woman was savagely beaten and left for dead, and the *shiftah* dragged Osman with them to the hills.

There, they cut his throat.

News came of his severed arm, sent to the village as a warning of what would happen to others if they spied. The rest of his body was never found: no doubt it was left for the hyena that roam each night the barren Ethiopian hills.

At the time of writing, Aidan is flying to Moyale to find out more and to help and comfort his family.

When I return to Kenya, and we have obtained the necessary passes, we will go back together to look for Osman's remains. Almost certainly, it will be impossible to find them. But we owe it to Osman to try.

As a faithful Muslim, he must be buried with his head towards Mecca. Perhaps, this will be the time that I will call in my favour to the Abagatha of the Borana people. His support and protection will be invaluable in that harsh land, riddled with rebels, where human life has no value at all.

I shall write that account in due course. Now is the time to mourn. Sadness descends on us when we remember our companion in adventure, and reflect on the meaning of his violent and barbarous death. Our consolation is that Osman died bravely, and for one of his faith this represents certainty to be in the paradise of Allah, in the company of heroes.

As it is written.

London, March 1998

Glossary

Akili	Clever
Amani	Peace
Arufu	Smell
Asante	Thank you
Askari	Guard, night-watchman
Ausiku	Night
Aya, ayah	Nanny, maid
Bahary	Sea
Bara bara	Road, track
Bhang, bhangi	Marijuana
Bila shaka	Without fail
Boma	Enclosure made of thorn bushes
Buibui	Traditional dress for Muslim women
Bule	Lava
Bunduki	Gun
Bwana	Mr, husband
Carissa	Wild jasmine with edible berries
Changaa	Strong, addictive, illegal local alcoholic brew
Chapati	Local pancake
Chinja	Slaughter
Chui	Leopard
Chùmvi	Salt
Daho	Wooden Arabic sailing boat
Damu	Blood
Debe	Can
Demu	Dam, pond, lake
Dengi	Deadly type of fever, similar to malaria
Desturi	Custom
Duka	Local shop
Faru, kifaru	Rhinoceros
Fundi	Craftsman, artisan
Galla	Camel, in Somali

Gamia	Camel, in Swahili
Gored gored	Ethiopian dish of raw meat
Habari?	What news?
Hapa	Here
Hapana	No
Ingine	Another
Inshallah	If God wills
Jabia	Traditional Arab method of gathering rainwater in underground tanks
Jambo	Hello
Jangili	Poachers, bandits
Jembe	Spade
Jua	Sun
Kaburi	Grave
Kamili	Exactly
Kanga	Loose garment for women, local printed wrap for women
Kaputi	Coat
Kaskazi	Ocean breeze
Kesha	Tomorrow
Kifaru, faru	Rhinoceros
Kikapu	Basket
Kikoi	Loose garment for men, local striped wrap for men
Kisima	Spring
Kitanda	Bed
Kopje	Small hill
Kufuata	To follow
Kubwa	Large
Kufa	Die, dead
Kufuata	To follow
Kuja	Come
Kulia	To the right
Kumbuka	Remember
Kuni	Firewood
Kuona, ku ona	To see
Kwaheri	Goodbye
Kwenda	Go
Lelechwa	Wild sage

Lugga	Dried-up river bed, small valley, ditch
Mabati	Corrugated iron
Madafu	Fresh coconut
Maho	Small traditional wooden boat
Maji	Water
Makena	The One Who Smiles
Makuti	Thatch made of palm leaves, thatch for roofing
Manyatta	Dwellings for a family group built within an enclosure, traditional living enclosure, Kraal
Maramoja	At once
Marati	Trough
Mashua	Boat
Mawe	Rock
Mawingo	Clouds
Mboga	Vegetable, cabbage
Mbogani	Small clearing, small plain
Mbusi	Goat
Mchanga	Sand
Melango	Door, gap
Memsaab	Madam
Menanda	Cattle dip
Meno	Teeth
Mianzi	Bamboo
Mimba	Pregnant
Miti	Tree
Mlima	Hill
Mnyama	Animal
Muchanga	Soil, sand
Muganga	Witch-doctor, wizard
Mugongo	Back
Mugu	Feet, tracks
Mugumu	Sacred fig tree
Muivi	Thief
Mukignei	Evergreen shrub of euclea
Murram	Red soil
Musungu, Wasungu	European(s)
Mutamayo	Wild olive
Mutandiko	Saddle
Mutoto	Child

Mvua	Rain
Mwaka	Year
Mzee	Old man
Napenda	Like
Nashukuru	I am grateful
Ndege	Bird, aeroplane
Ndiyo	Yes
Ndofu, ndovu	Elephant
Ndongo/Ndonga	Ochre
Ngamia	Camel
Ngombe	Cattle
Ngurue	Pig
Ni shauri ya Mungu	It is the will of God
Njau	Calf
Nungu nungu	Porcupine
Nussu	Half
Nyama	Meat
Nyapara	Headman
Nyayo	Track
Nyeupe	White
Nyeusi	Black
Nyoka	Snake
Nyota	Stars
Nyukundu	Red
Nyumba	House, room
Panga	Machete
Papa	Shark
Pembe	Horn
Pete	Ring
Poriti	Mangrove poles
Pole	Sorry
Pole-pole	Slowly
Posho	Maizemeal
Rafiki	Friend
Ringa	Slasher
Roho	Heart
Rondavel	Metal prefabricated octagonal huts
Rudi	Return
Rungu	Club, stick

Saidia	Help
Samaki	Fish
Samaki ya mugu	Crayfish, fish with legs
Samousas	Local fried pasty with spiced meat or vegetable filling
Shamba	Small farm, field, small holding
Shammah	Traditional Arab man's headscarf
Shauri	Problem
Shiftah	Bandits
Shimu	Hole
Shuka	Loincloth, wrap, shawl
Simba	Lion
Sokho	Open local market
Sufuria	Pot, pan
Taabu	Problem
Tafa dali, Tafadhali	Please
Tandarua	Tarpaulin
Tangu	Since
Tena	Again
Tope	Mud
Upupa	Hoopoe
Wachungai	Cattle herders
Wanawake	Women
Wasungu, Musungu	European(s)
Watu	People
Wazee	Elders, wise men
Zaidi	More